Essentials

A Corvette Collector's Guide
1956-1967

Chuck Brigermann

Copyright Chuck Brigermann 1992. All rights reserved. Except for short quotations used for reviews, reproduction, by any means, without permission, is prohibited. Not responsible for inaccuracies pertaining to any, or all of the information contained in this publication, and no liability is assumed in connection with the use of this data or specific details.

We realize that some words, model names, and designations mentioned herein are the property of the trademark holder. We use them for identification purposes only. This is not an official publication.

Distributed by Motorbooks International. Books are available at discounts in bulk quantity for industrial or sales promotional use. For details write to:
    Special Sales Manager
    Motorbooks International
    PO Box 2
    729 Prospect Avenue
    Osceola, WI 54020

Motorbooks International is a certified trademark, registered with the United States Patent Office.

Responsible comments, and verifiable information are always welcome, and should be directed to the publisher.

ISBN 0-89709-207-4

INTRODUCTION

Not unlike many other things of value, authenticity and originality are primary considerations when dealing with older Corvettes. This is the basis for the book. Regardless of the level of, or reason for your interest, a correct original car is the best value, and safest investment. This book offers a wealth of information for the novice or the serious collector.

The book covers 1956-1967. The prime years. Each year has it's own section, and each section contains all of the categories for fast, easy reference. At the beginning of each section will be definitive line art drawings of each model, along with calendars to verify build date information. Other drawings will be included, where necessary, to insure understanding and comprehension of the facts.

As with any endeavor of this magnitude, differences of opinion will exist, and mistakes will be made. Any responsible criticism, or additional information is always welcome.

Hope you enjoy it!

Chuck Brigermann

ACKNOWLEDGEMENTS

Tom Ambrose
Paul Baker
Dick Benton
Tom Christmann
Ken Clemens
Roger Fellows
John Grata
Ed Gurdjian
Ken Hanna
Gary Hodges
Scott Kohn
Bob Kunz
Jeff Little
Tom Luedeke
Michael Malstrom
Jerry Pings
John Pirkle
Tim Pope
Dennis Portka
Darrell Sheppard
Dale Smith
Brian Tilles
Roger Williams

A special thanks to General Motors for their information and assistance - and, of course, the Corvette.

Special recognition to Tom Anjiras for the beautiful line art, Dwight Waggener for the cover design, and Bill Seitz for the cover art.

TABLE OF CONTENTS

Chapter	Page
1956	1
1957	17
1958	37
1959	57
1960	77
1961	97
1962	117
1963	137
1964	157
1965	177
1966	201
1967	223

CORVETTE '56

1955

```
        JANUARY                    FEBRUARY
S   M   T   W   T   F   S     S   M   T   W   T   F   S
                        1               1   2   3   4   5
2   3   4   5   6   7   8     6   7   8   9  10  11  12
9  10  11  12  13  14  15    13  14  15  16  17  18  19
16 17  18  19  20  21  22    20  21  22  23  24  25  26
23 24  25  26  27  28  29    27  28
30 31

         MARCH                       APRIL
S   M   T   W   T   F   S     S   M   T   W   T   F   S
            1   2   3   4   5                         1   2
6   7   8   9  10  11  12     3   4   5   6   7   8   9
13 14  15  16  17  18  19    10  11  12  13  14  15  16
20 21  22  23  24  25  26    17  18  19  20  21  22  23
27 28  29  30  31            24  25  26  27  28  29  30

          MAY                        JUNE
S   M   T   W   T   F   S     S   M   T   W   T   F   S
1   2   3   4   5   6   7                 1   2   3   4
8   9  10  11  12  13  14     5   6   7   8   9  10  11
15 16  17  18  19  20  21    12  13  14  15  16  17  18
22 23  24  25  26  27  28    19  20  21  22  23  24  25
29 30  31                    26  27  28  29  30

          JULY                      AUGUST
S   M   T   W   T   F   S     S   M   T   W   T   F   S
                    1   2         1   2   3   4   5   6
3   4   5   6   7   8   9     7   8   9  10  11  12  13
10 11  12  13  14  15  16    14  15  16  17  18  19  20
17 18  19  20  21  22  23    21  22  23  24  25  26  27
24 25  26  27  28  29  30    28  29  30  31
31

       SEPTEMBER                    OCTOBER
S   M   T   W   T   F   S     S   M   T   W   T   F   S
                1   2   3                             1
4   5   6   7   8   9  10     2   3   4   5   6   7   8
11 12  13  14  15  16  17     9  10  11  12  13  14  15
18 19  20  21  22  23  24    16  17  18  19  20  21  22
25 26  27  28  29  30        23  24  25  26  27  28  29
                             30  31

        NOVEMBER                   DECEMBER
S   M   T   W   T   F   S     S   M   T   W   T   F   S
        1   2   3   4   5                     1   2   3
6   7   8   9  10  11  12     4   5   6   7   8   9  10
13 14  15  16  17  18  19    11  12  13  14  15  16  17
20 21  22  23  24  25  26    18  19  20  21  22  23  24
27 28  29  30                25  26  27  28  29  30  31
```

1956

```
        JANUARY                    FEBRUARY
S   M   T   W   T   F   S     S   M   T   W   T   F   S
1   2   3   4   5   6   7               1   2   3   4
8   9  10  11  12  13  14     5   6   7   8   9  10  11
15 16  17  18  19  20  21    12  13  14  15  16  17  18
22 23  24  25  26  27  28    19  20  21  22  23  24  25
29 30  31                    26  27  28  29

         MARCH                       APRIL
S   M   T   W   T   F   S     S   M   T   W   T   F   S
                    1   2   3     1   2   3   4   5   6   7
4   5   6   7   8   9  10     8   9  10  11  12  13  14
11 12  13  14  15  16  17    15  16  17  18  19  20  21
18 19  20  21  22  23  24    22  23  24  25  26  27  28
25 26  27  28  29  30  31    29  30

          MAY                        JUNE
S   M   T   W   T   F   S     S   M   T   W   T   F   S
        1   2   3   4   5                             1   2
6   7   8   9  10  11  12     3   4   5   6   7   8   9
13 14  15  16  17  18  19    10  11  12  13  14  15  16
20 21  22  23  24  25  26    17  18  19  20  21  22  23
27 28  29  30  31            24  25  26  27  28  29  30

          JULY                      AUGUST
S   M   T   W   T   F   S     S   M   T   W   T   F   S
1   2   3   4   5   6   7                 1   2   3   4
8   9  10  11  12  13  14     5   6   7   8   9  10  11
15 16  17  18  19  20  21    12  13  14  15  16  17  18
22 23  24  25  26  27  28    19  20  21  22  23  24  25
29 30  31                    26  27  28  29  30  31

       SEPTEMBER                    OCTOBER
S   M   T   W   T   F   S     S   M   T   W   T   F   S
                        1         1   2   3   4   5   6
2   3   4   5   6   7   8     7   8   9  10  11  12  13
9  10  11  12  13  14  15    14  15  16  17  18  19  20
16 17  18  19  20  21  22    21  22  23  24  25  26  27
23 24  25  26  27  28  29    28  29  30  31
30

        NOVEMBER                   DECEMBER
S   M   T   W   T   F   S     S   M   T   W   T   F   S
                1   2   3                                 1
4   5   6   7   8   9  10     2   3   4   5   6   7   8
11 12  13  14  15  16  17     9  10  11  12  13  14  15
18 19  20  21  22  23  24    16  17  18  19  20  21  22
25 26  27  28  29  30        23  24  25  26  27  28  29
                             30  31
```

1956 Vehicle Identification Number (VIN) E56S001001 to E56S004467, total vehicles 3,467. Plate attaches with ordinary Phillips screws to driver's side door jamb, just below upper hinge.
E=Corvette
56=model year
S=St. Louis
Six digit sequential serial number starting at 001001

AIR CLEANERS

Most single four-barrels used a louvered (four rows) chrome air cleaner 11" in diameter. The base must have a depression in order to clear the choke. Two four-barrel cars used a similar 6 3/4" unit made of aluminum. Neither unit had a serviceable filter element. Some early single four-barrel cars used the 6 3/4" style.

AIR CONDITIONING

Not Available

ANTENNAS

Assembly part # 3749220. Three section mast with a plastic cap containing raised ridges at the base, a washer and nut (not visible installed), a plastic spacer and gasket. Mast will have a ball at the top and may or may not have a ring at the top of the lower sections.

BALANCERS

Narrow, 6" diameter balancer with the pulley riveted directly to it.

BATTERIES

1890587 tar top. Year/month and factory of origin is stamped in the top. Cover above the tar on top should be 3 1/2 to 4 inches wide. Caps are all yellow with Delco on the outside, visible casting line/flashing on the bottom, inside. Case side should say "Delco Original Equipment Line" and should be unpainted (warranty replacements were painted). The characters 2SMR53 should be stamped in the side and top of the battery.

BATTERY CABLES

Negative cable is an uninsulated woven strap with an N on the bolt style terminal connection. Positive cable is black vinyl covered with a P on the bolt style terminal connection.

BELLHOUSINGS

3704922 early, and 3733365, both cast iron and date coded.

BRAKE MASTER CYLINDERS

Undated. Part # 5454480 located in the metal plug in the end of the cylinder. Casting # 5450233 located on the driver's side just above the 1" bore size identification. Filler cap was metal, not plastic, and contains USE GMC BRAKE FLUID around the top.

CARBURETORS

Model numbers are embossed on a brass tag attached under one of the carburetor top screws. The date is also shown on this tag in the form month/year, using A=Jan, B=Feb, etc., an additional digit indicated factory production information.

All 210 HP used WCFB 2366S(first design) or 2366SA with the choke mounted at the top, and the choke line entering through the side of the housing, not the cover. The main body was cast 0-953, the top 6-1098 for 2366S and 6-1151 for 2366SA. This carburetor did not have a secondary counterweight.

All 225 HP and 240 HP used WCFB 2419S front and 2362S rear. The front carburetor does not use a choke. The choke arrangement for the rear is the same as the single four. Both use secondary counterweights, and a main body cast 0-049. Tops could be cast 6-1114, 6-1122(both early usage with 3728735 intake) or 6-1151 and 6-1156.

CARPET

Extremely short, tight loop, Daytona weave with foam backing. Sectionally sewn to fit the individual areas and contours of the interior. Has an almost mat-like appearance rather than loop carpet.

COILS

1115083 or 1115091, with 083 or 091 in raised characters on the case.

CONVERTIBLE TOP

All tops had the rear window sewn into place. The rear window contained the Vinylite name and trademark, the AS 6 identification, cleaning instructions, and a date code hot stamped into the outside surface. The rear bow should match the body contour of the top compartment lid, and uses a mohair rear seal. It also requires the use of filler weatherstrips that are somewhat shorter than the 1961-1962 type. The straps attach to the rear bow with a tack strip, not a clamp. The header should be the 56-58 style with a rounded shoe and plate attachment to the side frame. Front hold down latches should have a short handle (1 7/8" across top), not the later 2 1/2" replacement handle. The base of the latch is rounded compared to the tapered angle of the hardtop latch.

CRANKCASE VENTS

The system consists of the oil filler cap and the road draft tube. The road draft tube had a few minor bends between the top and the spark plug grommet holder, the remaining lower area had only one distinct bend between two straight sections. See the separate oil cap listing.

DISTRIBUTORS

Part number and date (year/month/day) stamped on a black and aluminum tag which is riveted to the distributor housing. Part numbers should be 1110872 or 1110879.

EMBLEMS

Nose & Trunk

ENGINE BLOCKS

Casting # 3720991. All engines were 265 cubic inches. Casting number is located on the driver's side rear flange where the bellhousing attaches. The date is located directly opposite on the passenger side rear flange. They are dated month/day/year, with a single digit year representation. Double digits indicate Tonawanda manufacture and are not correct for Corvette. The month is indicated by a consecutive alphabetic character, ie January = A, February = B, etc. A code indicating the assembly location, date, and engine characteristics was stamped on a pad located on the passenger side of the motor. It is just below the cylinder head, next to the water pump. The first character should be F for the Flint assembly plant. The numeric characters represent the date, and the last two characters indicate the engine type as shown in the chart that follows. In 1956 the pad also contained a serial number that was not related to the vehicle serial number.

- FK = STD 210HP, 4 barrel, automatic transmission
- GV = STD 210HP, 4 barrel, 3 speed
- FG = 469 225HP, 2 four barrels, automatic transmission
- GR = 469 225HP, 2 four barrels, 3 speed
- GU = 449 240HP, 2 four barrels, 3 speed, special cam

EXHAUST MANIFOLDS

Dated month and day only for correct Flint manifolds. Casting # 3725563 early (about the first 720 cars), both sides use the same manifold with a two bolt exhaust pipe flange. Casting # 3731557 left, and 3731558 right, with a three bolt flange, for the balance of the year.

FUEL FILTERS

All cars used a glass sediment bowl style of filter. A 'stone' filter element was used and the bowl was held in place by a wire bail. The cast alloy top contained a large AC in the center of a distinct dome.

FUEL INJECTION

Not Available

FUEL PUMPS

The AC logo is cast directly in the top of the pump. The four digit Delco number is stamped in the edge of the mounting flange and should be 4262 early, 4346 for the remainder of the year. Both pumps are screwed together assemblies with the early pump using short, 3/4" screws with one diaphragm, and the 4346 using a much longer screw with a lot of threads exposed between two diaphragms.

GAUGES

Fuel/Temp - Battery/Oil combinations on the small gauges contain convex lenses with dot indicators on an inner plastic face. Pointers are rounded and a chrome gauge name plate with a circle in the center covers the lower half of the gauge face. All tachs are 6000 rpm without a red line marker and have a rev counter, an inner face with dots, another inner face with numbers, and an outer convex lens. Most clocks have a square back and are made by Westclox (stamped on the back), some 1955 "bullet back" Delco clocks were used early in the year. Speedometers are 140 MPH having an inner face with dots, and a flat outer face with numbers.

GENERATORS

1102043/30 Amp, all engines.

The model number and a date code appear on a red/aluminum tag attached to the generator. The year, month, day, are listed as serial number on the tag, and the month appears as an alphabetic character with A=Jan, B=Feb, etc. All units use a one piece fan-pulley assembly with a 3/8" belt groove and a 3 3/8" diameter pulley. Tach drive assembly is attached to the rear.

GLASS

All laminated glass was LOF Safety Plate, and along with the logo, was designated as such by AS1 for windshields and AS2 for side glass. Side and rear Plexiglas was AS4. A two character code (month-year) also appeared indicating the date of manufacture. January=N, February=M, March=L, April=K, May=J, June=I, July=H, August=T, September=E, October=F, November=C, December=V. The letter X indicates 1955, the letter V 1956. Tinted glass was not used.

GRILLE

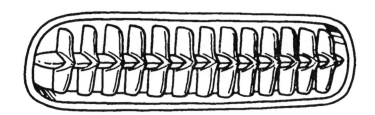

HARDTOP

Should have a 1" wide aluminum header trim strip. The lower rear stainless trim consists of three pieces joined at the back of each quarter window by a Y shaped escutcheon. All AS-4 Plexiglas windows are dated month/year. Front hold down latches should have a short handle(1 7/8" across top), not the later 2 1/2" replacement handle. The base of the latch is a tapered angle compared to the rounded soft top latch. Headliner material should be the same color and waffle pattern as the seats, etc.

HEADS

Casting #3725306 for 210 and 225HP with two bolt exhaust manifolds, 3731762 for 225 and 240HP with three bolt exhaust manifolds. Dated month/day/year, with a single digit year representation. Double digits indicate Tonawanda manufacture and are not correct for Corvette. The month is indicated by a consecutive alphabetic character, ie January = A, February = B, etc. Symbol at the end of the head is a partial indication of the application.

HORNS

The last three digits of the part number are stamped into the horn. The high note is 1999760, the low note 1999759. The production date is not stamped. Mounting brackets were removable, not welded to the horn, and somewhat obscure the part number. Unlike any of the later design Corvette horns, these are the large dome style.

HORN RELAYS

1116913 for all cars. The characters '913', for the last three digits of the part number, and '12VN', for 12 volt, negative ground, are stamped in the mounting flange. The cover is stamped with Delco Remy.

HUBCAPS

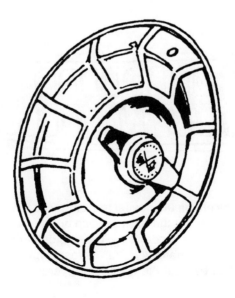

IGNITION SHIELDING

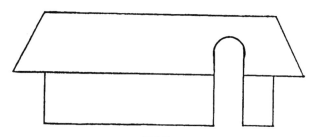

TOP

The first shields were chrome plated steel with a round hole in the top. In the spring of 1956 the top was changed to stainless steel, and the round hole was changed to a slot. The front and rear skirt are straight across, and only contain an opening for the accelerator linkage. The front skirt is visibly shorter than the rear skirt. There should be a depression in the upper rear skirt near the middle. All rear attachments are through holes, not slots.

VERTICALS

All left shields should have the longer, triangular bottom, and are made from chrome plated steel. The right shield lower mounting should be through a hole.

HORIZONTALS

Initial design shielding also doubled as heat shielding. Late in the year, a single piece of shielding was attached to each side purely for radio interference suppression. Both sides have the smaller exhaust manifold cut-outs, and no reinforcement bar.

INTAKE MANIFOLDS

Dated month/day/year. Casting numbers are located on the top surface. Cast iron manifold date codes are also located on the top surface, and are therefore visible with the manifold installed. Correct manifolds should have a single digit year representation. Aluminum manifold date codes are cast into the bottom surface and are not visible unless the manifold is removed from the engine. Cast iron 210HP manifold was 3735448. Initial usage for the 225HP aluminum manifold was 3728735. This was changed to 3731394 for both the 225 and 240HP application.

JACK

Correct scissor jack has an identification number, SJ4653, cast directly into the top of the cast iron base. Jack arms and saddle are stamped steel. Jack screw has fine threads and the handle facility is an oblong slot in the end.

KNOBS

LAMPS

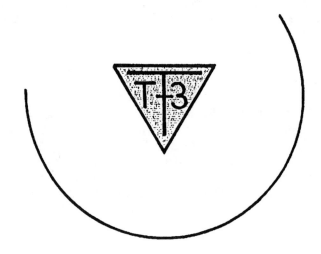

MIRRORS

    Outside rearview mirror was stamped 'GUIDE Y-50' on the mirror back at the ball stud. This ball should be centered in the back of the mirror. The glass was not dated, and considered non replaceable since the later style snap ring access was not provided.

    Inside mirrors had a base containing a thumbscrew that allowed you to raise and lower the mirror along with the usual swivel adjustment. The mirror glass was not dated.

OIL CAPS

Non vented to SN 3000

OPTIONS

		Quantity	Price
2934	Base convertible	3,467	$3,120.00
Incl.	265, 210HP	276	NC
Incl.	3 Speed		NC
Incl.	Black folding top	103	NC
101	Heater	2,953	123.65
102	Wonderbar AM radio	2,717	198.90
107	Parking brake alarm	2,685	5.40
108	Courtesy light	2,775	8.65
109	Windshield washers	2,815	11.85
290	Whitewall tires	2,815	32.20
313	Powerglide	715	188.30
419	Hardtop	1,447	215.20
419	Hardtop without soft top	629	NC
426	Power windows	547	64.60
440	Two tone paint	1,259	19.40
449	Special cam	111	188.30
469	265, 225HP engine	3,080	188.30
470	White soft top	1,840	NC
470	Beige soft top	895	NC
473	Power folding top	2,682	107.60

PAINT/TRIM

1956-1962 Corvettes did not have a tag that provided paint and trim code information.

EXTERIOR	COVE	INTERIOR	SOFT TOP
Onyx Black	Silver	Red	Black or White
Venetian Red	Beige	Red	Beige or White
Aztec Copper	Beige	Beige	Beige or White
Polo White	Silver	Red	Black or White
Cascade Green	Beige	Beige	Beige or White
Artic Blue	Silver	Red or Beige	Beige or White

Cove colors shown were optional, body color standard. These were factory recommended combinations.

RADIATORS

All radiators were copper and contained an identification tag soldered to the top tank. This tag contained the part number 3133689, and the date of manufacture represented by a double digit year code and a single alphabetic month code (Jan=A, Feb=B, etc.). Fan shroud was the wide, four piece style.

RADIATOR CAPS

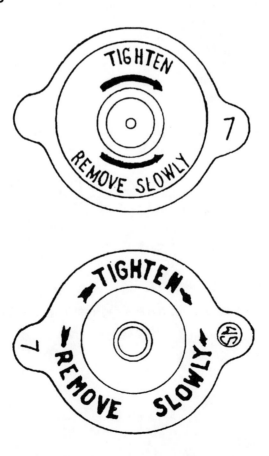

RADIOS

Model number 3711897, 9 tube Wonder Bar. Touch bar contains no writing, tuning indicator is white, and the dial face contains the Chevrolet bow tie. A paper tag on the outside of one of the radio covers contains the model number.

REAR AXLE

Code and build date (month/day) are located on the right, front of the carrier.
CODES
AE .... 3.55, Powerglide
AH .... 3.70, 3 speed
AD .... 3.27, 3 speed
AJ .... 4.11, 3 speed

SEATS

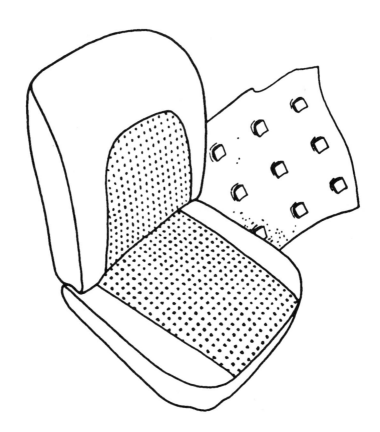

SHIFTERS

The three speed shifter and the automatic transmission shifter have a similar overall appearance even including the mounting bracket. Both utilized a round, thin, chrome plated shift lever. Automatic shift pattern was L D N R P front to back.

SPARK PLUG WIRES

Distributor boot is 180 degree, black. Spark plug boot is 90 degree, orange. Wire is 7MM black with the imprint RADIO GM TVRS, undated.

STARTERS

1107627 was used on all cars. The model number and a date code appear on a red/aluminum tag attached to the generator. The year, month, day, are listed as serial number on the tag, and the month appears as an alphabetic character with A=Jan, B=Feb, etc.

STEERING WHEELS

Color matching (red or beige) wheels were plastic with metal spokes. Outer plastic circumference should be free of grind marks found on reproduction wheels. Finishing marks on the spokes should run the same direction as the rim on the face of the spoke and perpendicular to the rim on the back. The amount of simulated grain, or lack of it, appearing in the plastic of original wheels is a point of controversy. Replacement GM wheels have plenty, but some very original cars have little or none. The idea that the grain has worn away is unlikely since it is typically missing from all areas of the wheel, even areas with little hand contact.

THERMOSTAT HOUSINGS

Engines with a cast iron intake manifold used a curved, cast iron housing. Engines with an aluminum intake manifold used an angled, cast iron housing. It did not have a lip at the end of the neck to insure hose sealing.

TIRES

Firestone 'Deluxe Champion', US Royal 'Air Ride', or B.F. Goodrich 'Silvertown' 6.70x15, tubeless. If whitewall equipped, the width should be 2" to 2 11/16" nominal. DOT (department of transportation) information did not appear in the sidewall of original tires.

TRANSMISSIONS

Four speed was not available in 1956. Build information is stamped on the upper rear of the case, passenger side for three speeds, lower rear of the case, passenger side for Powerglide. Build information is in the form: location/month/day/shift (D or N). S=Saginaw (three speed), C=Cleveland (Powerglide). The three speed must have an unvented tail housing (cast 946) with a shifter mounting position and cut outs at the rear bushing.

VALVE COVERS

All valve cover mounting holes were closer together on top than they were on the bottom. The 210HP engine used a stamped steel cover with CHEVROLET in raised script commonly called "staggered holes". The remaining engines used a cast aluminum cover with CORVETTE in raised script and nine fins running the length of the cover. After the first several months of production two notches were added to the inside lip for intake manifold clearance.

VOLTAGE REGULATORS

1119001 was used on all cars. The cover was stamped with Delco Remy and was attached with slotted cap screws. The part number and date, along with '12VN' for 12 volt negative ground, are stamped in the mounting flange. The date is shown as year/month, with the month represented by an alphabetic character, ie A=Jan, B=Feb, etc.

WATER PUMPS

Casting # 3704911, undated. Should have an unmachined boss at the top that is used for a bypass fitting in other applications.

WIPER ARMS & BLADES

All cars used a bright stainless, Trico brand, with rubber insert. The Trico name may not always appear on the parts. The rubber inserts have a series of raised dots running the length of both sides. The blade construction consists of a main bridge piece that is somewhat shorter than the overall length of the blade. Attached to this are three links that hold the rubber insert. There are also two hard rubber blocks at each end of the blade.

WIPER MOTORS

Early 5047223 unverified, 5047924 all others.

Number is stamped in the front cover which has a metal slide (as opposed to plastic). A ground wire should not be riveted to the rear cover arm. A 1 1/2" long metal support stand belongs on the lower armature housing bolt. Windshield washers were optional equipment.

CORVETTE '57

1956

```
        JANUARY                    FEBRUARY
S   M   T   W   T   F   S      S   M   T   W   T   F   S
1   2   3   4   5   6   7                  1   2   3   4
8   9  10  11  12  13  14      5   6   7   8   9  10  11
15 16  17  18  19  20  21     12  13  14  15  16  17  18
22 23  24  25  26  27  28     19  20  21  22  23  24  25
29 30  31                     26  27  28  29

         MARCH                       APRIL
S   M   T   W   T   F   S      S   M   T   W   T   F   S
                1   2   3      1   2   3   4   5   6   7
4   5   6   7   8   9  10      8   9  10  11  12  13  14
11 12  13  14  15  16  17     15  16  17  18  19  20  21
18 19  20  21  22  23  24     22  23  24  25  26  27  28
25 26  27  28  29  30  31     29  30

          MAY                         JUNE
S   M   T   W   T   F   S      S   M   T   W   T   F   S
        1   2   3   4   5                              1   2
6   7   8   9  10  11  12      3   4   5   6   7   8   9
13 14  15  16  17  18  19     10  11  12  13  14  15  16
20 21  22  23  24  25  26     17  18  19  20  21  22  23
27 28  29  30  31             24  25  26  27  28  29  30

          JULY                       AUGUST
S   M   T   W   T   F   S      S   M   T   W   T   F   S
1   2   3   4   5   6   7                  1   2   3   4
8   9  10  11  12  13  14      5   6   7   8   9  10  11
15 16  17  18  19  20  21     12  13  14  15  16  17  18
22 23  24  25  26  27  28     19  20  21  22  23  24  25
29 30  31                     26  27  28  29  30  31

        SEPTEMBER                   OCTOBER
S   M   T   W   T   F   S      S   M   T   W   T   F   S
                            1      1   2   3   4   5   6
2   3   4   5   6   7   8      7   8   9  10  11  12  13
9  10  11  12  13  14  15     14  15  16  17  18  19  20
16 17  18  19  20  21  22     21  22  23  24  25  26  27
23 24  25  26  27  28  29     28  29  30  31
30

        NOVEMBER                   DECEMBER
S   M   T   W   T   F   S      S   M   T   W   T   F   S
                1   2   3                              1
4   5   6   7   8   9  10      2   3   4   5   6   7   8
11 12  13  14  15  16  17      9  10  11  12  13  14  15
18 19  20  21  22  23  24     16  17  18  19  20  21  22
25 26  27  28  29  30         23  24  25  26  27  28  29
                              30  31
```

1957

```
        JANUARY                    FEBRUARY
S   M   T   W   T   F   S      S   M   T   W   T   F   S
        1   2   3   4   5                              1   2
6   7   8   9  10  11  12      3   4   5   6   7   8   9
13 14  15  16  17  18  19     10  11  12  13  14  15  16
20 21  22  23  24  25  26     17  18  19  20  21  22  23
27 28  29  30  31             24  25  26  27  28

         MARCH                       APRIL
S   M   T   W   T   F   S      S   M   T   W   T   F   S
                        1   2      1   2   3   4   5   6
3   4   5   6   7   8   9      7   8   9  10  11  12  13
10 11  12  13  14  15  16     14  15  16  17  18  19  20
17 18  19  20  21  22  23     21  22  23  24  25  26  27
24 25  26  27  28  29  30     28  29  30
31

          MAY                         JUNE
S   M   T   W   T   F   S      S   M   T   W   T   F   S
            1   2   3   4                                  1
5   6   7   8   9  10  11      2   3   4   5   6   7   8
12 13  14  15  16  17  18      9  10  11  12  13  14  15
19 20  21  22  23  24  25     16  17  18  19  20  21  22
26 27  28  29  30  31         23  24  25  26  27  28  29
                              30

          JULY                       AUGUST
S   M   T   W   T   F   S      S   M   T   W   T   F   S
    1   2   3   4   5   6                      1   2   3
7   8   9  10  11  12  13      4   5   6   7   8   9  10
14 15  16  17  18  19  20     11  12  13  14  15  16  17
21 22  23  24  25  26  27     18  19  20  21  22  23  24
28 29  30  31                 25  26  27  28  29  30  31

        SEPTEMBER                   OCTOBER
S   M   T   W   T   F   S      S   M   T   W   T   F   S
1   2   3   4   5   6   7              1   2   3   4   5
8   9  10  11  12  13  14      6   7   8   9  10  11  12
15 16  17  18  19  20  21     13  14  15  16  17  18  19
22 23  24  25  26  27  28     20  21  22  23  24  25  26
29 30                         27  28  29  30  31

        NOVEMBER                   DECEMBER
S   M   T   W   T   F   S      S   M   T   W   T   F   S
                        1   2      1   2   3   4   5   6   7
3   4   5   6   7   8   9      8   9  10  11  12  13  14
10 11  12  13  14  15  16     15  16  17  18  19  20  21
17 18  19  20  21  22  23     22  23  24  25  26  27  28
24 25  26  27  28  29  30     29  30  31
```

1957 Corvette

1957 Vehicle Identification Number (VIN) E57S100001 to E57S106339, total vehicles 6,339. Plate attaches with ordinary Phillips screws to driver's side door jamb, just below upper hinge.
E=Corvette
57=model year
S=St. Louis
Six digit sequential serial number starting at 100001

AIR CLEANERS

Single four-barrel used a louvered (four rows) chrome air cleaner about 11" in diameter. The base must have a depression in order to clear the choke. Two four-barrel cars used a similar 6 3/4" unit made of aluminum. Neither unit had a serviceable filter element.

Two fuel injection air cleaners were used, the first looked like the carburetor style with more rows of louvers, and also did not have a serviceable element. The second was actually a screened paper element (A56C with blue ends, replacements had black ends) with a chrome lid held together by the wing nut. This was used about the last 1/3 of production.

AIR CONDITIONING

Not Available

ANTENNAS

Assembly part # 3749220. Three section mast with a plastic cap containing raised ridges at the base, a washer and nut (not visible installed), a plastic spacer and gasket. Mast will have a ball at the top and may or may not have a ring at the top of the lower sections.

BALANCERS

Narrow, 6" diameter balancer with the pulley riveted directly to it was used on all but 283HP engines. The 283HP engines used basically the same balancer with a bolt on pulley.

BATTERIES

1980458 tar top. Year/month and factory of origin is stamped in the top. Cover above the tar on top should be 3 1/2 to 4 inches wide. Caps are all yellow with Delco on the outside, visible casting line/flashing on the bottom, inside. Case side should say "Delco Original Equipment Line" and should be unpainted (warranty replacements were painted). The characters 2SMR53 should be stamped in the side and CAT NO 458 on top of the battery.

BATTERY CABLES

Negative cable is an uninsulated woven strap with an N on the bolt style terminal connection. Positive cable is black vinyl covered with a P on the bolt style terminal connection.

BELLHOUSINGS

3733365, cast iron and date coded.

BRAKE MASTER CYLINDERS

Undated. Part # 5454480 located in the metal plug in the end of the cylinder. Casting # 5456022 in equal size characters located on the driver's side just above the 1" bore size identification. Filler cap was metal, not plastic, and contains USE GMC BRAKE FLUID around the top.

CARBURETORS

Model numbers are embossed on a brass tag attached under one of the carburetor top screws. The date is also shown on this tag in the form month/year, using A=Jan, B=Feb, etc., an additional digit indicated factory production information.

The 220 HP used 2655S. The 2655S has the choke mounted at the top, but the choke line is attached to the center of the choke cover. This is true even though the side entry fitting is still present. A secondary counterweight is used. The main body is cast 0-108, the top 6-1271 or 6-1299.

The 245 HP used WCFB 2419S front and 2362S rear, very early only. The front carburetor does not use a choke. The choke arrangement for the rear is the same as the 2627S below. Both front and rear use secondary counterweights, and a main body cast 0-049. Tops could be cast 6-1098, 6-1114, 6-1122, 6-1151, or 6-1156. The majority of the 245 HP applications and all 270 HP used WCFB 2626S front, WCFB 2627S rear. Both contain secondary counterweights, but no idle air screws. A choke was mounted on the top of the rear carburetor only, and the line was attached at the side, not on the cover. Main bodies are cast 0-049, tops 6-1299 or 6-1049.

CARPET

Extremely short, tight loop, Daytona weave with foam backing. Sectionally sewn to fit the individual areas and contours of the interior. Has an almost mat-like appearance.

COILS

1115091 for carbureted cars, with 091 in raised characters on the case. 1115107 with fuel injection, 107 in raised characters on the case.

CONVERTIBLE TOP

All tops had the rear window sewn into place. The rear window contained the Vinylite name and trademark, the AS 6 identification, cleaning instructions, and a date code hot stamped into the outside surface. Although blueprints show a rear window with angled sides, wider at the bottom than at the top, original factory tops came with parallel sides and a rectangular appearance. The rear bow should match the body contour of the top compartment lid, and uses a mohair rear seal. It also requires the use of filler weatherstrips that are somewhat shorter than the 1961-1962 type. The straps attach to the rear bow with a tack strip, not a clamp. The header should be the 56-58 style with a rounded shoe and plate attachment to the side frame. Front hold down latches should have a short handle (1 7/8" across top), not the later 2 1/2" replacement handle. The base of the latch is rounded compared to the tapered angle of the hardtop latch.

CRANKCASE VENTS

The system consists of the oil filler cap and the road draft tube. The road draft tube had a few minor bends between the top and the spark plug grommet holder, the remaining lower area had only one distinct bend between two straight sections. See the separate oil cap listing.

DISTRIBUTORS

Part number and date (year/month/day) stamped on a black and aluminum tag which is riveted to the distributor housing. Part numbers should be 1110891 for all except fuel injection. Fuel injection cars used 1110906 with automatic transmission, 1110889 (early) or 1110905 with manual transmission, except air box equipped cars. Air box equipped cars used 1110908 which had a tach drive facility. Some of these distributors were used on late non-air box applications with the tach drive plugged.

EMBLEMS

Nose & Trunk

Side, FI Only

Side, Trunk

ENGINE BLOCKS

Casting # 3731548. All engines were 283 cubic inches. Casting number is located on the driver's side rear flange where the bellhousing attaches. The date is located directly opposite on the passenger side rear flange. They are dated month/day/year, with a single digit year representation. Double digits indicate Tonawanda manufacture and are not correct for Corvette. The month is indicated by a consecutive alphabetic character, ie January = A, February = B, etc. A code indicating the assembly location, date, and engine characteristics was stamped on a pad located on the passenger side of the motor. It is just below the cylinder head, next to the water pump. The first character should be F for the Flint assembly plant. The numeric characters represent the date, and the last two characters indicate the engine type as shown in the chart that follows. In 1957 the pad did not contain a serial number.

    EF = STD 220HP, 4 barrel, manual transmission
    FH = STD 220HP, 4 barrel, automatic transmission
    EH = 469 245HP, 2 four barrels, manual transmission
    FG = 469 245HP, 2 four barrels, automatic transmission
    EM = 579 250HP, fuel injection, manual transmission
    FK = 579 250HP, fuel injection, automatic transmission
    EG = 469 270HP, 2 four barrels, special cam, manual
    EL = 579 283HP, fuel injection, special cam, manual

EXHAUST MANIFOLDS

Dated month and day only for correct Flint manifolds. Casting # 3733975 left, 3733976 right.

FUEL FILTERS

All carbureted cars used a glass sediment bowl style of filter. A 'stone' filter element was used and the bowl was held in place by a wire bail. The cast alloy top contained a large AC in the center of a distinct dome.

Fuel injected cars used a canister style filter with a removable element similar to an oil filter. The canister was held in place by a stamped steel bracket attached to the intake manifold. Both fuel lines connected to the flat, top portion of the assembly, and the bosses at these connections were initially round. At approximately mid production the bosses were cast square for wrench support. The canister housings were smooth for the very early cars, then changed to a ribbed design. A weep hole should be present in the bottom of the canister.

FUEL INJECTION

Four different units were used in 1957. All units had a ribbed plenum that contained a black/aluminum tag showing the part number and a serial number on the driver's side (some of the very first units do not have a tag, the number is stamped on the rear of the plenum). The serial number has no relationship to any other number. All cold start enrichment was provided by a micro switch activated solenoid, therefore no cranking signal valves were used in 1957. This arrangement was assisted by an electric thermostatic coil mounted on the air meter. A conventional choke with it's associated butterfly was not used.

7014360   All engines. Early units had white buttons in the ends of the nozzle blocks, a coast valve on the fuel meter, and a rear vent line, along with the unique dual spider fuel distribution. All units used short nozzles that could not be disassembled. Restrictor T fitting was used on top of the

	main diaphragm, with each line running to opposite sides of the air meter. The air cleaner adaptor had three line connections, two from the nozzle balance tubes, and one from the forward balance tube that goes to the fuel meter. Snowflake was cast into the plenum top. Air meter stamped 7014361, cast 7014054 Fuel meter stamped 7014362, cast 7014312
7014520	All engines. Only the forward balance tube ran to the air cleaner adaptor even though the provision for three lines still existed. The two unused tubes were covered with red caps. The nozzle balance tube was connected directly to the air meter. Two lines ran from the main diaphragm T fitting to opposite sides of the air meter. Enrichment diaphragm line runs to the front of the plenum. Snowflake was cast into the plenum top. During production a switch was made to a serviceable nozzle. Air meter stamped 7014521, cast 7014054 early, or 7014388 Fuel meter stamped 7014522, cast 7014312
7014800	250HP. Snowflake was not cast into plenum top. A single line air cleaner adaptor was used and was connected to the fuel meter via the forward balance tube. Two lines ran from the main diaphragm T fitting to opposite sides of the air meter. Air meter stamped 7014801, cast 7014804 Fuel meter stamped 7014802, cast 7014312
7014960	283HP. Same basic appearance as 7014800 Air meter stamped 7014801, cast 7014804 Fuel meter stamped 7014962, cast 7014312

FUEL PUMPS

The AC logo is cast directly in the top of the pump. The four digit Delco number is stamped in the edge of the mounting flange and should be 4346. Pumps are screwed together assemblies using a long screw with a lot of threads exposed between two diaphragms.

GAUGES

Fuel/Temp - Battery/Oil combinations on the small gauges contain convex lenses with dot indicators on an inner plastic face. Pointers are rounded and a chrome gauge name plate with a circle in the center covers the lower half of the gauge face. All dash mounted tachs are 6000 rpm without a red line marker and have a rev counter, an inner face with dots, another inner face with numbers, and an outer convex lens. Some of the air box equipped cars had an 8000 RPM column mounted tach. Clocks have a square back and are made by Westclox (stamped on the back). Speedometers are 140 MPH having an inner face with dots, and a flat outer face with numbers.

1957 Corvette

GENERATORS

⊙1102043/30 Amp, all except FI with an air box
1102059/30 Amp, FI cars with an air box

The model number and a date code appear on a red/aluminum tag attached to the generator. The year, month, day, are listed as serial number on the tag, and the month appears as an alphabetic character with A=Jan, B=Feb, etc. All units use a one piece fan-pulley assembly. All 220HP had a 3/8" belt groove and a 3 3/8" diameter pulley. All other 1102043 applications used a 3 5/8" pulley with a 1/2" belt groove. The 1102059 used a 4" pulley with a 1/2" belt groove. Tach drive assembly is attached to the rear on all but 1102059.

GLASS

All laminated glass was LOF Safety Plate and, along with the logo, was designated as such by AS1 for windshields and AS2 for side glass. Side and rear Plexiglas was AS4. A two character code (month-year) also appeared indicating the date of manufacture. January=N, February=M, March=L, April=K, May=J, June=I, July=H, August=T, September=E, October=F, November=C, December=V. The letter V indicates 1956, the letter T 1957. Tinted glass was not used.

GRILLE

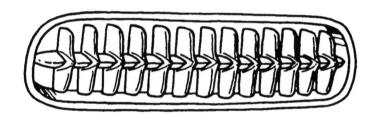

HARDTOP

Should have a 2" wide, two piece, stainless header trim strip with a small cap covering the center joint. The lower rear stainless trim consists of three pieces joined at the back of each quarter window by a Y shaped escutcheon. All AS-4 Plexiglas windows are dated month/year. Front hold down latches should have a short handle (1 7/8" across top), not the later 2 1/2" replacement handle. The base of the latch is a tapered angle compared to the rounded soft top latch. Headliner material should be the same color and waffle pattern as the seats, etc.

HEADS

Casting # 3740997 for 220, 245, 250 and 270HP, 3731539 for 283HP. Dated month/day/year, with a single digit year representation. Double digits indicate Tonawanda manufacture and are not correct for Corvette. The month is indicated by a consecutive alphabetic character, ie January = A, February = B, etc. Symbol at the end of the head is a partial indication of the application.

HORNS

The last three digits of the part number are stamped into the horn. The high note is 9000340, the low note 9000339. The production date is not stamped. Mounting brackets were removable, not welded to the horn.

HORN RELAYS

1116913 for all cars. The characters '913', for the last three digits of the part number, and '12VN', for 12 volt, negative ground, are stamped in the mounting flange. The cover is stamped with Delco Remy.

HUBCAPS

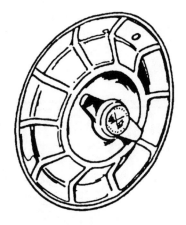

Standard

RPO

IGNITION SHIELDING

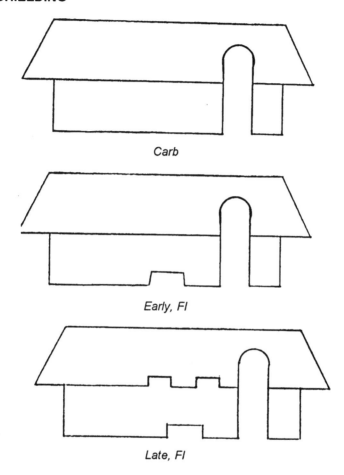

Carb

Early, FI

Late, FI

TOP

All shields were stainless steel and had a front skirt that was visibly shorter than the rear. There should be a depression in the upper rear skirt near the middle. All rear attachments are through holes, not slots. For carbureted cars the front and rear skirt are straight across, and only contain an opening for the accelerator linkage. For cars with fuel injection there is a bulge in the front skirt to accommodate the coil, and until late in the production year, a single notch appeared in the rear skirt. During late production two notches were also added to the front skirt.

VERTICALS

All left shields should have the longer, triangular bottom, until about May when the triangular portion was removed. Both are made from chrome plated steel. The right shield lower mounting should be through a hole.

HORIZONTALS
A single piece of shielding was used on each side. Both sides have the smaller exhaust manifold cut-outs, and no reinforcement bar.

INTAKE MANIFOLDS
Dated month/day/year. Casting numbers are located on the top surface. Cast iron manifold date codes are also located on the top surface, and are therefore visible with the manifold installed. Correct manifolds should have a single digit year representation. Aluminum manifold date codes are cast into the bottom surface and are not visible unless the manifold is removed from the engine. Cast iron 220HP manifold was 3731398. The aluminum manifold was 3739653 for both the 245 and 270HP applications.

JACK
Correct scissor jack has an identification number, SJ4653, cast directly into the top of the cast iron base. Jack arms and saddle are stamped steel. Jack screw has fine threads and the handle facility is an oblong slot in the end.

KNOBS

LAMPS

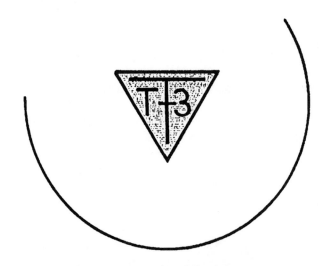

MIRRORS

Outside rearview mirror was stamped 'GUIDE Y-50' on the mirror back at the ball stud. This ball should be centered in the back of the mirror. The glass was not dated, and considered non replaceable since the later style snap ring access was not provided.

Inside mirrors had a base containing a threaded hole and a locknut that allowed you to raise and lower the mirror along with the usual swivel adjustment. The stainless back contained either two or three rivets, and may or may not have an S logo. The mirror glass was not dated.

OIL CAPS

Solid lifter engines were non vented

OPTIONS

		Quantity	Price
2934	Base convertible	6,339	$3,176.32
Incl.	283, 220HP	1,633	NC
Incl.	3 Speed	4,282	NC
Incl.	Black folding top	1,180	NC
101	Heater	5,373	118.40
102	Wonderbar AM radio	3,635	199.10
107	Parking brake alarm	1,873	5.40
108	Courtesy light	2,489	8.65
109	Windshield washers	2,555	11.85
276	15X5 1/2 Wheels	51	15.10
290	Whitewall tires	5,019	31.60
313	Powerglide	1,393	188.30
419	Hardtop	3,124	215.20
419	Hardtop without soft top	1,571	NC
426	Power windows	379	59.20
440	Two tone paint	3,026	19.40
469A	283, 245HP engine	2,045	150.65
469C	283, 270HP engine	1,021	182.95
470	White soft top	2,461	NC

470	Beige soft top	1,127	NC
473	Power folding top	1,336	139.90
579A	283, 250HP engine	182	484.20
579B	283, 283HP engine	713	484.20
579C	283, 250HP engine, automatic trans	102	484.20
579E	283, 283HP engine, with air box	43	726.30
677	Positraction rear, 3:703	27	48.45
678	Positraction rear, 4:11	1,772	48.45
679	Positraction rear, 4:56	6	48.45
684	HD Brakes & suspension	51	780.10
685	4 Speed transmission	664	188.30

DEALER AVAILABLE OPTIONS
Spotlight

PAINT/TRIM

1956-1962 Corvettes did not have a tag that provided paint and trim code information.

EXTERIOR	COVE	INTERIOR	SOFT TOP
Onyx Black	Silver	Red:Beige	Black:White:Beige
Venetian Red	Beige	Red:Beige	Beige:White:Black
Aztec Copper	Beige	Beige	Beige:White
Polo White	Silver	Red:Beige	Black:White:Beige
Cascade Green	Beige	Beige	Beige:White:Black
Arctic Blue	Silver	Red:Beige	Beige:White:Black
Inca Silver	Ivory	Red:Beige	Black:White

Cove colors shown were optional, body color standard.
These were factory recommended combinations.

PRODUCTION FIGURES

	MONTHLY	CUMULATIVE
October	580	580
November	490	1070
December	NA	NA
January	NA	NA
February	NA	NA
March	NA	NA
April	NA	NA
May	NA	4331
June	593	4924
July	660	5584
August	645	6229
September	110	6339

RADIATORS

All radiators were copper and contained an identification tag soldered to the top tank. This tag contained the part number 3133689, and the date of manufacture represented by a double digit year code and a single alphabetic month code (Jan=A, Feb=B, etc.). The fan shroud was the wide, four piece style.

RADIATOR CAPS

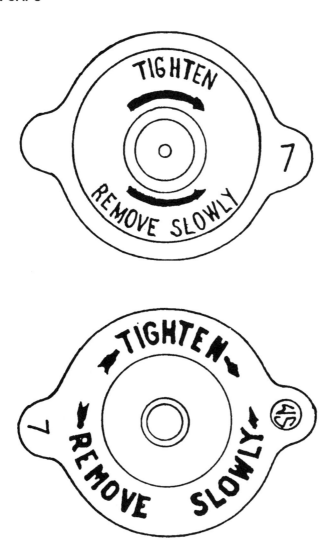

RADIOS

Model number 3725156, 10 tube Wonder Bar. Touch bar contains 'WONDER BAR', tuning indicator is white, and the dial face contains the Chevrolet bow tie. A paper tag on the outside of one of the radio covers contains the model number.

REAR AXLE

Code and build date (month/day) are located on the right, front of the carrier.
CODES
AE .... 3.36, Open, Powerglide
AH .... 3.70, Open, manual
AJ .... 4.11, Open, manual
AN .... 3.70, Positraction, manual
AP .... 4.11, Positraction, manual
AQ .... 4.56, Positraction, manual
AS .... 3.70, Positraction, manual, HD brakes & suspension
AT .... 4.11, Positraction, manual, HD brakes & suspension
AU .... 4.56, Positraction, manual, HD brakes & suspension

SEATS

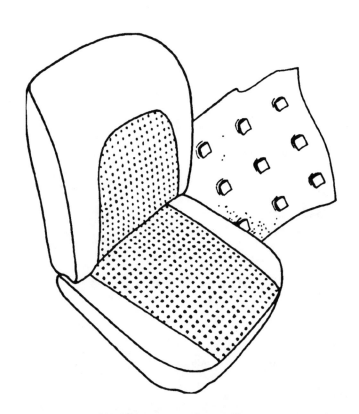

SHIFTERS

The three speed, four speed, and the automatic transmission shifter have a similar overall appearance, even including the mounting bracket. All utilized a round, thin, chrome plated shift lever. Automatic shift pattern was L D N R P front to back. The four speed shifter was unique to 1957 incorporating a clip for reverse control and a large, round, unslotted nut on the shifter tower shaft. There should not be a reverse lockout or anti rattle spring.

SPARK PLUG WIRES

Distributor boot is 180 degree on carbureted cars, 90 degree on fuel injected cars, black. Spark plug boot is 90 degree, orange. Wire is 7MM black with the imprint RADIO GM TVRS, undated.

STARTERS

1107664 was used on all cars. The date and model number were stamped directly into the starter housing. The date is shown as year, month, day, and the month appears as an alphabetic character with A=Jan, B=Feb, etc. The solenoid bakelite should contain Delco Remy.

STEERING WHEELS

Color matching (red or beige) wheels were plastic with metal spokes. Outer plastic circumference should be free of grind marks found on reproduction wheels. Finishing marks on the spokes should run the same direction as the rim on the face of the spoke and perpendicular to the rim on the back. The amount of simulated grain, or lack of it, appearing in the plastic of original wheels is a point of controversy. Replacement GM wheels have plenty, but some very original cars have little or none. The idea that the grain has worn away is unlikely since it is typically missing from all areas of the wheel, even areas with little hand contact.

THERMOSTAT HOUSINGS

Engines with a cast iron intake manifold used a curved, cast iron housing. Engines with an aluminum intake manifold used an angled, cast iron housing, on all except late production which could be aluminum. The cast iron housing did not have a lip at the end of the neck to insure radiator hose sealing. The aluminum housing, number 3837223, should not have metal saver notches around the stud bosses. This area should be flat.

TIRES

Firestone 'Deluxe Champion', US Royal 'Air Ride', or B.F. Goodrich 'Silvertown' 6.70x15, tubeless. If whitewall equipped, the width should be 2" to 2 11/16" nominal. DOT (department of transportation) information did not appear in the sidewall of original tires.

TRANSMISSIONS

The Borg Warner four speed became available late (around May) in the production year. Some of the earliest units (first 20) were undated. The aluminum tail housings have two half moon cuts at the rear bushing and only one shifter mounting position (3 holes), SPAC should be cast below the date. The side cover is dated inside and should not contain any webbing. The main case is cast iron and has T10-1 and month/day cast into the passenger side. A build code is stamped on the upper rear, driver's side, unmachined part of the case. The code is in the form: location/month/day/year/shift, W=Warner Gear, month=Jan=A, Feb=B, etc., shift=1,2,3. Vehicle serial numbers are not stamped on these transmissions.

Build information is stamped on the upper rear of the case, passenger side for three speeds, lower rear of the case, passenger side for Powerglide. Build information is in the form: location/month/day/shift (D or N). S=Saginaw (three speed), C=Cleveland (Powerglide). The three speed must have a shifter mounting position on the tail housing (cast 450), and should have cut outs at the rear bushing.

VALVE COVERS

All valve cover mounting holes were closer together on top than they were on the bottom. The 220HP engine used a stamped steel cover with CHEVROLET in raised script. The remaining engines used a cast aluminum cover with CORVETTE in raised script about 1/32" high. Two notches appeared on the inside lip for intake manifold clearance. Early production used a cover with nine fins running the length of the cover. Starting around late February a cover with seven fins was phased in. The covers with nine fins were used until the supply was exhausted around June.

VOLTAGE REGULATORS

1119001 was used on all cars. The cover was stamped with Delco Remy and was attached with slotted cap screws. The part number and date, along with '12VN' for 12 volt negative ground, are stamped in the mounting flange. The date is shown as year/month, with the month represented by an alphabetic character, ie A=Jan, B=Feb, etc.

WATER PUMPS

Casting # 3736493, undated.

WIPER ARMS & BLADES

All cars used a bright stainless, Trico brand, with rubber insert. The Trico name may not always appear on the parts. The rubber inserts have a series of raised dots running the length of both sides. The blade construction consists of a main bridge piece that is somewhat shorter than the overall length of the blade. Attached to this are three links that hold the rubber insert. There are also two hard rubber blocks at each end of the blade.

WIPER MOTORS

5047984 until about serial 1400, 5047991 for the balance of the year. The number is stamped in the front cover. The 5047991 motor incorporated the slow park feature. There should not be a ground wire riveted to the rear cover arm, and there may or may not be a 1 1/2" metal support stand attached to the lower armature housing bolt. Windshield washers were optional equipment.

CORVETTE '58

1957

JANUARY
S	M	T	W	T	F	S
		1	2	3	4	5
6	7	8	9	10	11	12
13	14	15	16	17	18	19
20	21	22	23	24	25	26
27	28	29	30	31		

FEBRUARY
S	M	T	W	T	F	S
					1	2
3	4	5	6	7	8	9
10	11	12	13	14	15	16
17	18	19	20	21	22	23
24	25	26	27	28		

MARCH
S	M	T	W	T	F	S
					1	2
3	4	5	6	7	8	9
10	11	12	13	14	15	16
17	18	19	20	21	22	23
24	25	26	27	28	29	30
31						

APRIL
S	M	T	W	T	F	S
	1	2	3	4	5	6
7	8	9	10	11	12	13
14	15	16	17	18	19	20
21	22	23	24	25	26	27
28	29	30				

MAY
S	M	T	W	T	F	S
			1	2	3	4
5	6	7	8	9	10	11
12	13	14	15	16	17	18
19	20	21	22	23	24	25
26	27	28	29	30	31	

JUNE
S	M	T	W	T	F	S
						1
2	3	4	5	6	7	8
9	10	11	12	13	14	15
16	17	18	19	20	21	22
23	24	25	26	27	28	29
30						

JULY
S	M	T	W	T	F	S
	1	2	3	4	5	6
7	8	9	10	11	12	13
14	15	16	17	18	19	20
21	22	23	24	25	26	27
28	29	30	31			

AUGUST
S	M	T	W	T	F	S
				1	2	3
4	5	6	7	8	9	10
11	12	13	14	15	16	17
18	19	20	21	22	23	24
25	26	27	28	29	30	31

SEPTEMBER
S	M	T	W	T	F	S
1	2	3	4	5	6	7
8	9	10	11	12	13	14
15	16	17	18	19	20	21
22	23	24	25	26	27	28
29	30					

OCTOBER
S	M	T	W	T	F	S
		1	2	3	4	5
6	7	8	9	10	11	12
13	14	15	16	17	18	19
20	21	22	23	24	25	26
27	28	29	30	31		

NOVEMBER
S	M	T	W	T	F	S
					1	2
3	4	5	6	7	8	9
10	11	12	13	14	15	16
17	18	19	20	21	22	23
24	25	26	27	28	29	30

DECEMBER
S	M	T	W	T	F	S
1	2	3	4	5	6	7
8	9	10	11	12	13	14
15	16	17	18	19	20	21
22	23	24	25	26	27	28
29	30	31				

1958

JANUARY
S	M	T	W	T	F	S
			1	2	3	4
5	6	7	8	9	10	11
12	13	14	15	16	17	18
19	20	21	22	23	24	25
26	27	28	29	30	31	

FEBRUARY
S	M	T	W	T	F	S
						1
2	3	4	5	6	7	8
9	10	11	12	13	14	15
16	17	18	19	20	21	22
23	24	25	26	27	28	

MARCH
S	M	T	W	T	F	S
						1
2	3	4	5	6	7	8
9	10	11	12	13	14	15
16	17	18	19	20	21	22
23	24	25	26	27	28	29
30	31					

APRIL
S	M	T	W	T	F	S
		1	2	3	4	5
6	7	8	9	10	11	12
13	14	15	16	17	18	19
20	21	22	23	24	25	26
27	28	29	30			

MAY
S	M	T	W	T	F	S
				1	2	3
4	5	6	7	8	9	10
11	12	13	14	15	16	17
18	19	20	21	22	23	24
25	26	27	28	29	30	31

JUNE
S	M	T	W	T	F	S
1	2	3	4	5	6	7
8	9	10	11	12	13	14
15	16	17	18	19	20	21
22	23	24	25	26	27	28
29	30					

JULY
S	M	T	W	T	F	S
		1	2	3	4	5
6	7	8	9	10	11	12
13	14	15	16	17	18	19
20	21	22	23	24	25	26
27	28	29	30	31		

AUGUST
S	M	T	W	T	F	S
					1	2
3	4	5	6	7	8	9
10	11	12	13	14	15	16
17	18	19	20	21	22	23
24	25	26	27	28	29	30
31						

SEPTEMBER
S	M	T	W	T	F	S
	1	2	3	4	5	6
7	8	9	10	11	12	13
14	15	16	17	18	19	20
21	22	23	24	25	26	27
28	29	30				

OCTOBER
S	M	T	W	T	F	S
			1	2	3	4
5	6	7	8	9	10	11
12	13	14	15	16	17	18
19	20	21	22	23	24	25
26	27	28	29	30	31	

NOVEMBER
S	M	T	W	T	F	S
						1
2	3	4	5	6	7	8
9	10	11	12	13	14	15
16	17	18	19	20	21	22
23	24	25	26	27	28	29
30						

DECEMBER
S	M	T	W	T	F	S
	1	2	3	4	5	6
7	8	9	10	11	12	13
14	15	16	17	18	19	20
21	22	23	24	25	26	27
28	29	30	31			

1958 Vehicle Identification Number (VIN) J58S100001 to J58S109168, total vehicles 9,168. Plate attaches with ordinary Phillips screws to driver's side door jamb, just below upper hinge.
J=Corvette
58=model year
S=St. Louis
Six digit sequential serial number starting at 100001

AIR CLEANERS

Both the two four barrel and the single four barrel used a similar louvered air cleaner approximately 14 inches in diameter. There were three rows of louvers in the lightly polished aluminum top. The top was crimped to the base, therefore neither unit had a serviceable filter. The tops were held in place with wing nuts at the carburetor centers. The wing nuts sat in depressions in the top, and the two four barrel unit had an additional depression in the center.

The fuel injection air cleaner is a two piece unit held together with a wing nut. This provided access to the serviceable filter element. The units are all metal and have a large X stamped in the lid that extends out to the very edge of the air cleaner. The units did not bolt directly to the engine, but to the driver's side inner fender.

AIR CONDITIONING

Not Available

ANTENNAS

Assembly part # 3749220. Three section mast with a plastic cap containing raised ridges at the base, a washer and nut (not visible installed), a plastic spacer and gasket. Mast will have a ball at the top and may or may not have a ring at the top of the lower sections.

BALANCERS

Narrow, 6" diameter balancer with the pulley bolted directly to it.

BATTERIES

1980458 tar top. Year/month and factory of origin is stamped in the top. Cover above the tar on top should be 3 1/2 to 4 inches wide. Caps are all yellow with Delco on the outside, visible casting line/flashing on the bottom, inside. Case side should say "Delco Original Equipment Line" and should be unpainted (warranty replacements were painted). The characters 2SMR53 should be stamped in the side and CAT NO 458 on top of the battery.

BATTERY CABLES

Negative cable is an uninsulated woven strap with an N on the bolt style terminal connection. Positive cable is black vinyl covered with a P on the bolt style terminal connection.

BELLHOUSINGS

3733365, cast iron and date coded.

BRAKE MASTER CYLINDERS

Undated. Part # 5454480 located in the metal plug in the end of the cylinder. Casting # 5456022 in equal size characters located on the driver's side just above the 1" bore size identification. Filler cap was metal, not plastic, and contains USE GMC BRAKE FLUID around the top.

CARBURETORS

Model numbers are embossed on a brass tag attached under one of the carburetor top screws. The date is also shown on this tag in the form month/year, using A=Jan, B=Feb, etc., an additional digit indicated factory production information.
All 230 HP used WCFB 2669S. The 2669S has the choke mounted at the top, and the choke line is attached to the center of the choke cover. A secondary counterweight is used. The main body is cast 0-208, the top 6-1354 or 6-1208.
All 245 HP used WCFB 2626S front, WCFB 2627S rear. When 270 HP

production resumed in January all used WCFB 2613S front and 2614S rear. All contain secondary counterweights, but no idle air screws. A choke was mounted on the top of the rear carburetor only, and the line was attached at the side, not on the cover. Main bodies are cast 0-049, tops 6-1299 or 0-1049.

CARPET

Extremely short, tight loop, Daytona weave with foam backing. Sectionally sewn to fit the individual areas and contours of the interior. Has an almost mat-like appearance.

COILS

1115091 for carbureted cars, with 091 in raised characters on the case. 1115107 with fuel injection, 107 in raised characters on the case.

CONVERTIBLE TOP

All tops had the rear window sewn into place. The rear window contained the Vinylite name and trademark, the AS 6 identification, cleaning instructions, and a date code hot stamped into the outside surface. The rear bow should match the body contour of the top compartment lid, and uses a mohair rear seal. It also requires the use of filler weatherstrips that are somewhat shorter than the 1961-1962 type. The straps attach to the rear bow with a tack strip, not a clamp. The header should be the 56-58 style with a rounded shoe and plate attachment to the side frame. Front hold down latches should have a short handle (1 7/8" across top), not the later 2 1/2" replacement handle. The base of the latch is rounded compared to the tapered angle of the hardtop latch.

CRANKCASE VENTS

The system consists of the oil filler cap and the road draft tube. The road draft tube had a few minor bends between the top and the spark plug grommet holder, the remaining lower area had only one distinct bend between two straight sections. Some late cars may have used the 1959 style vent. See the separate oil cap listing.

DISTRIBUTORS

The model used with the 230HP base engine contains a metal tag that wraps around the neck of the distributor below the cap. This contains the date and model number 1110890. All 245 and 270HP cars used model number 1110891 that, along with the date, was stamped on a black and aluminum tag riveted to the distributor housing. Early 290HP fuel injection cars used model 1110908 with the same type of tag used on the 270HP described above. The remaining fuel injection cars used models with a cross shaft access cover. The date and model number were stamped in the cover. The 250HP used 1110906 early, then 1110915. The 290HP used

1110914. Fuel injection distributors all had a facility for driving the tach cable. All models were dated year/month/day.

EMBLEMS

Nose & Trunk (Gold Letters)

Side

Side

ENGINE BLOCKS

Casting numbers 3737739 or 3756519. The 3756519 casting appeared around mid year. All engines were 283 cubic inches. Casting number is located on the driver's side rear flange where the bellhousing attaches. The date is located directly opposite on the passenger side rear flange. They are dated month/day/year, with a single digit year representation. Double digits indicate Tonawanda manufacture and are not correct for Corvette. The month is indicated by a consecutive alphabetic character, ie January = A, February = B, etc. A code indicating the assembly location, date, and engine characteristics was stamped on a pad located on the

passenger side of the motor. It is just below the cylinder head, next to the water pump. The first character should be F for the Flint assembly plant. The numeric characters represent the date, and the last two characters indicate the engine type as shown in the chart that follows. In 1958 the pad did not contain a serial number.
    CQ = STD 230HP, 4 barrel, manual transmission
    DG = STD 230HP, 4 barrel, automatic transmission
    CT = 469 245HP, 2 four barrels, manual transmission
    DJ = 469 245HP, 2 four barrels, automatic transmission
    CR = 579 250HP, fuel injection, manual transmission
    DH = 579 250HP, fuel injection, automatic transmission
    CU = 469 270HP, 2 four barrels, special cam, manual
    CS = 579 290HP, fuel injection, special cam, manual

EXHAUST MANIFOLDS

Dated month and day only for correct Flint manifolds. Casting # 3749965 left, 3750556 right.

FUEL FILTERS

All carbureted cars used a glass sediment bowl style of filter. A 'stone' filter element was used and the bowl was held in place by a wire bail. The cast alloy top contained a large AC in the center of a distinct dome for most of 1958 production. A flatter top with a smaller AC was used for the balance of the year.

Fuel injected cars used a canister style filter with a removable element similar to an oil filter. The canister was held in place by a stamped steel bracket attached to the intake manifold. Both fuel lines connected to the flat, top portion of the assembly, and the bosses at these connections were cast square for wrench support. The canister housings were a ribbed design. A weep hole should be present in the bottom of the canister.

FUEL INJECTION

Five different units were used in 1958. All units had a ribbed plenum that contained a black/aluminum tag showing the part number and a serial number on the driver's side. The serial number has no relationship to any other number. A snowflake was not cast into the plenum top. Cold start enrichment was assisted by an electric thermostatic coil mounted on the air meter. See unit description for additional details. A conventional choke with it's associated butterfly was not used.

7014800   250HP. Cold start enrichment was provided by a micro switch activated solenoid, therefore no cranking signal valve was used. Forward balance tube ran from the fuel meter to the air cleaner adaptor. Two lines ran from the main diaphragm to opposite sides of the air meter. Enrichment diaphragm line ran across the front of the plenum to the air

	meter. Air meter stamped 7014801, cast 7014804 Fuel meter stamped 7014802, cast 7014312
7014800R	290HP. Essentially a reworked 7014520 with the same basic appearance as the 7014800. Air meter stamped 7014801, cast 7014388 Fuel meter stamped 7014802, cast 7014312
7014900	250HP. Cranking signal valve mounted in the plenum was part of the cold enrichment. Balance tube ran from the fuel meter to the air cleaner adaptor. Two lines ran from the main diaphragm, one to the air meter, and one to the cranking signal valve. Air meter stamped 7014901, cast 7014922 Fuel meter stamped 7014902, cast 7014889
7014900R	290HP. Merely a recalibrated 7014900 unit. Air meter stamped 7014901, cast 7014922 Fuel meter stamped 7014902, cast 7014889
7014960	290HP. Cold start enrichment was provided by a micro switch activated solenoid, therefore no cranking signal valve was used. Forward balance tube ran from the fuel meter to the air cleaner adaptor. Two lines ran from the main diaphragm to opposite sides of the air meter. Usage may be questionable for 1958. Air meter stamped 7014801, cast 7014804 Fuel meter stamped 7014962, cast 7014312

FUEL PUMPS

The AC logo is cast directly in the top of the pump. The four digit Delco number is stamped in the edge of the mounting flange and should be 4445 early, 4656 for the remainder of the year. Both pumps are single diaphragm, screwed together assemblies, using short, 3/4" screws.

GAUGES

Fuel/Temp - Battery/Oil combinations on the small gauges contain convex lenses with dot indicators on an inner plastic face. Pointers are rounded and a chrome gauge name plate with a circle in the center covers the lower half of the gauge face. All tachs in hydraulic lifter equipped cars are 6000 rpm with a red line marker and have a rev counter, an inner face with dots, another inner face with numbers, and an outer convex lens. Tachs in solid lifter equipped cars are 8000 RPM and do not have a rev counter. Clocks have a square back and are made by Westclox (stamped on the back). The F for the fast/slow indicator should be on the left. Speedometers are 160 MPH having an inner face with dots, and a flat outer face with numbers.

GENERATORS

1102043/30 Amp, all except 290HP
1102059/30 Amp, with 290HP

The model number and a date code appear on a red/aluminum tag attached to the generator. The year, month, day, are listed as serial number on the tag, and the month appears as an alphabetic character with A=Jan, B=Feb, etc. All units use a one piece fan-pulley assembly. All 1102043 applications used a 3 5/8" pulley with a 1/2" belt groove and a spacer between the fan and pulley. The 1102059 used a 4" pulley with a 1/2" belt groove. Tach drive assembly is attached to the rear on all but 1102059.

GLASS

All laminated glass was LOF Safety Plate and, along with the logo, was designated as such by AS1 for windshields and AS2 for side glass. Side and rear Plexiglas was AS4. A two character code (month-year) also appeared indicating the date of manufacture. January=N, February=M in 1957 X in 1958, March=L, April=K in 1957 G in 1958, May=J, June=I, July=H in 1957 U in 1958, August=T, September=E in 1957 A in 1958, October=F in 1957 Y in 1958, November=C, December=V. The letter T indicates 1957, the letter N 1958. Tinted glass was not used.

GRILLE

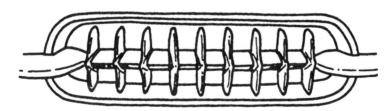

HARDTOP

Should have a 2" wide, two piece, stainless header trim strip with a small cap covering the center joint. The lower rear stainless trim consists of three pieces joined at the back of each quarter window by a Y shaped escutcheon. All AS-4 Plexiglas windows are dated month/year. Front hold down latches should have a short handle (1 7/8" across top), not the later 2 1/2" replacement handle. The base of the latch is a tapered angle compared to the rounded soft top latch. Headliner material should be the same color and pebble grain pattern as the seats, etc.

HEADS

Casting # 3748770. Dated month/day/year, with a single digit year representation. Double digits indicate Tonawanda manufacture and are not correct for

Corvette. The month is indicated by a consecutive alphabetic character, ie January = A, February = B, etc. Symbol at the end of the head is a partial indication of the application.

HORNS

The last three digits of the part number are stamped into the horn. The high note is 9000352, the low note 9000351. The production date is not stamped. Mounting brackets were welded to the horn, not bolted.

HORN RELAYS

1116913 for all cars. The characters '913', for the last three digits of the part number, and '12VN', for 12 volt, negative ground, are stamped in the mounting flange. The cover is stamped with Delco Remy.

HUBCAPS

Standard

RPO

IGNITION SHIELDING

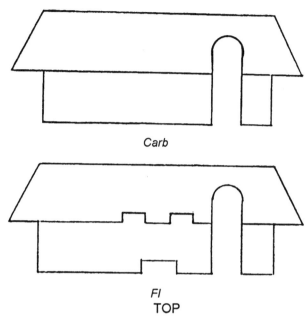

Carb

FI
TOP

All shields were stainless steel and had a front skirt that was visibly shorter than the rear. There should be a depression in the upper rear skirt near the middle. All rear attachments are through holes, not slots. For carbureted cars the front and rear skirt are straight across, and only contain an opening for the accelerator linkage. For cars with fuel injection there is a bulge in the front skirt to accommodate the coil, a single notch in the rear skirt, and two notches in the front skirt.

VERTICALS

All left shields should have the shorter, straight bottom. Both sides are made from chrome plated steel. The right shield lower mounting should be through a hole.

HORIZONTALS

A single piece of shielding was used on each side. Both sides have the smaller exhaust manifold cut-outs, and no reinforcement bar.

INTAKE MANIFOLDS

Dated month/day/year. Casting numbers are located on the top surface. Cast iron manifold date codes are also located on the top surface, and are therefore visible with the manifold installed. Correct manifolds should have a single digit year representation. Aluminum manifold date codes are cast into the bottom surface and are not visible unless the manifold is removed from the engine. Cast iron 230HP manifold was 3746829. The aluminum manifold was 3739653 for both the 245 and 270HP applications.

JACK

Correct scissor jack has an identification number, SJ4653, cast directly into the top of the cast iron base. Jack arms and saddle are stamped steel. Jack screw has fine threads and the handle facility is an oblong slot in the end.

KNOBS

LAMPS

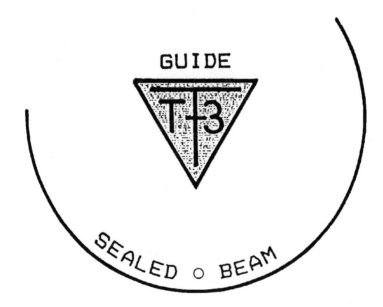

MIRRORS

Outside rearview mirror was stamped 'GUIDE Y-50' on the mirror back at the ball stud. This ball should be centered in the back of the mirror. The glass was not dated, and considered non replaceable since the later style snap ring access was not provided.

Inside mirrors had a base containing a threaded top where the mirror was attached and held in place with a locknut. The mirror was an exact rectangular shape with rounded corners and three rivets surrounding the ball stud in back. A 1/8" circle with an S in it appears 3/16" above the center rivet. The mirror glass was not dated.

OIL CAPS

Solid lifter engines were non vented

OPTIONS

		Quantity	Price
2934	Base convertible	9,168	$3,591.00
Incl.	283, 230HP	4,243	NC
Incl.	3 Speed	3,347	NC
Incl.	Black folding top	1,864	NC
101	Heater	8,014	96.85
102	Wonderbar AM radio	6,142	144.45
107	Parking brake alarm	2,883	5.40
108	Courtesy light	4,600	6.50
109	Windshield washers	2,834	16.15
276	15X5 1/2 Wheels	404	NC
290	Whitewall tires	7,428	31.55
313	Powerglide	2,057	188.30
419	Hardtop	3,392	215.20
419	Hardtop without soft top	2,215	NC
426	Power windows	649	59.20
440	Two tone paint	3,422	16.15
469	283, 245HP engine	2,436	150.65
469C	283, 270HP engine	978	182.95
470	White soft top	3,827	NC
470	Blue gray soft top	795	NC
473	Power folding top	1,090	139.90
579	283, 250HP engine	504	484.20
579D	283, 290HP engine	1,007	484.20
677	Positraction rear, 3:70	1,123	48.45
678	Positraction rear, 4:11	2,518	48.45
679	Positraction rear, 4:56	370	48.45
684	HD Brakes & suspension	144	780.10

685	4 Speed transmission	3,764	215.20

DEALER AVAILABLE OPTIONS
Tool Kit
Spotlight

PAINT/TRIM

1956-1962 Corvettes did not have a tag that provided paint and trim code information.

EXTERIOR	COVE	INTERIOR	SOFT TOP
Tuxedo Black	Silver	Red:Charcoal	Black:White
Signet Red	White	Red:Charcoal	White:Black
Regal Turquoise	White	Charcoal	Black:White
Snowcrest White	Silver	Red:Blue/Gray:Charcoal	Black:White:Blue/Gray
Charcoal	Silver	Red:Blue/Gray:Charcoal	White:Black
Silver Blue	Silver	Charcoal:Blue/Gray	Blue/Gray:White
Inca Silver	Black	Red:Charcoal	Black:White
Panama Yellow	White	Charcoal	Black:White

Cove colors shown were optional, body color standard.
These were factory recommended combinations.

PRODUCTION FIGURES

	MONTHLY	CUMULATIVE
October	486	486
November	957	1443
December	1068	2511
January	1166	3677
February	1112	4789
March	990	5779
April	765	6544
May	945	7489
June	703	8192
July	648	8840
August	328	9168

RADIATORS

All radiators were copper and contained an identification tag soldered to the top tank. This tag contained the part number 3141674, and the date of manufacture represented by a double digit year code and a single alphabetic month code (Jan=A, Feb=B, etc.). The fan shroud was the wide, four piece style.

RADIATOR CAPS

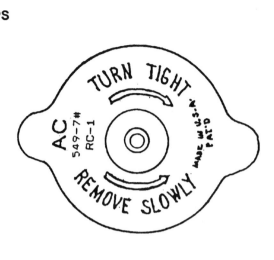

RADIOS

Model number 987730, 7 tube Wonder Bar. The face of the push buttons should be flat. A paper tag on the outside of one of the radio covers contains the model number.

REAR AXLE

Code and build date (month/day) are located on the right, front of the carrier.
CODES
AE .... 3.36, Open, Powerglide
AH .... 3.70, Open, manual
AJ .... 4.11, Open, manual
AN .... 3.70, Positraction, manual
AP .... 4.11, Positraction, manual
AQ .... 4.56, Positraction, manual
AS .... 3.70, Positraction, manual, HD brakes & suspension
AT .... 4.11, Positraction, manual, HD brakes & suspension
AU .... 4.56, Positraction, manual, HD brakes & suspension

SEATS

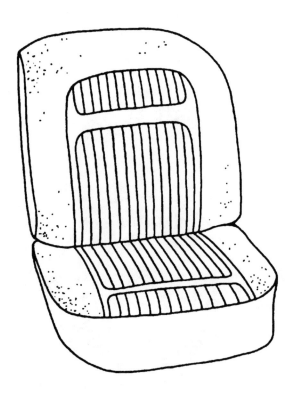

SHIFTERS

The three speed, four speed, and the automatic transmission shifter have a similar overall appearance even including the mounting bracket. All utilized a round, thin, chrome plated shift lever. Automatic shift pattern was L D N R P front to back. The four speed shifter was unique to 1958 incorporating a spring for gate tension and a large round nut that contained another spring for reverse control. Both were located on the shifter tower shaft. There should not be a reverse lockout or an anti rattle spring.

SPARK PLUG WIRES

Distributor boot is 180 degree on carbureted cars, 90 degree on fuel injected cars, black. Spark plug boot is 90 degree, orange. Wire is 7MM black with the imprint RADIO GM TVRS, undated.

STARTERS

1107664 was used on all cars. The date and model number were stamped directly into the starter housing. The date is shown as year, month, day, and the month appears as an alphabetic character with A=Jan, B=Feb, etc. The solenoid bakelite should contain Delco Remy.

STEERING WHEELS

Color matching (red, bluegray, or charcoal) wheels were plastic with metal spokes. Outer plastic circumference should be free of grind marks found on reproduction wheels. Finishing marks on the spokes should run the same direction as the rim on the face of the spoke and perpendicular to the rim on the back. The amount of simulated grain, or lack of it, appearing in the plastic of original wheels is a point of controversy. Replacement GM wheels have plenty, but some very original cars have little or none. The idea that the grain has worn away is unlikely since it is typically missing from all areas of the wheel, even areas with little hand contact.

THERMOSTAT HOUSINGS

Engines with a cast iron intake manifold used a curved, cast iron housing, number 3711268. Engines with an aluminum intake manifold used an angled, aluminum housing, number 3837223. The aluminum housing should not have metal saver notches around the stud bosses. This area should be flat.

TIRES

Firestone 'Deluxe Champion', US Royal 'Air Ride', or B.F. Goodrich 'Silvertown' 6.70x15, tubeless. If whitewall equipped, the width should be 2" to 2 9/16" nominal.

DOT (department of transportation) information did not appear in the sidewall of original tires.

TRANSMISSIONS

The Borg Warner four speed was available as an option. The aluminum tail housings have two half moon cuts at the rear bushing until mid February. There should be only one shifter mounting position (3 holes), and SPAC should be cast below the date (M castings used in 1959 production were first cast in May/June of 1958 and could show up). The side cover is dated inside and should not contain any webbing. The main case is cast iron and has T10-1 and month/day cast into the passenger side. A build code is stamped on the upper rear, driver's side, unmachined part of the case. The code is in the form: location/month/day/year/shift, W=Warner Gear, month=Jan=A, Feb=B, etc., shift=1,2,3. Vehicle serial numbers are not stamped on these transmissions.

Build information is stamped on the upper rear of the case, passenger side for three speeds, and on the rear flange of the governor cover for Powerglide. Build information is in the form: location/month/day/shift (D or N). S=Saginaw (three speed), C=Cleveland (Powerglide). The three speed must have a shifter mounting position on the tail housing.

VALVE COVERS

All valve cover mounting holes were closer together on top than they were on the bottom. The 230HP engine used a stamped steel cover with CHEVROLET in raised script.

The remaining engines used a cast aluminum cover with CORVETTE in raised script. Early production script was 1/32" high, later production 1/8". A casting flaw should not appear through the 'O' in CORVETTE. All covers had seven fins running the length of the cover. Two notches appeared on the inside lip for intake manifold clearance.

VOLTAGE REGULATORS

1119001 was used on all cars. The cover was stamped with Delco Remy and was attached with slotted cap screws. The part number and date, along with '12VN' for 12 volt negative ground, are stamped in the mounting flange. The date is shown as year/month, with the month represented by an alphabetic character, ie A=Jan, B=Feb, etc.

WATER PUMPS

Casting # 3736493, undated.

WIPER ARMS & BLADES

All cars used a bright stainless, Trico brand, with rubber insert. The Trico name may not always appear on the parts. The rubber inserts have a series of raised dots running the length of both sides. The blade construction consists of a main bridge piece that is somewhat shorter than the overall length of the blade. Attached to this are three links that hold the rubber insert. There are also two hard rubber blocks at each end of the blade.

WIPER MOTORS

5044266 stamped in the front cover. There should not be a ground wire riveted to the rear cover arm. There should be a 1" long metal support stand attached to the lower armature housing bolt. Windshield washers were optional equipment.

CORVETTE '59

1958

```
       JANUARY                    FEBRUARY
S  M  T  W  T  F  S        S  M  T  W  T  F  S
         1  2  3  4                          1
5  6  7  8  9 10 11        2  3  4  5  6  7  8
12 13 14 15 16 17 18       9 10 11 12 13 14 15
19 20 21 22 23 24 25      16 17 18 19 20 21 22
26 27 28 29 30 31         23 24 25 26 27 28

        MARCH                      APRIL
S  M  T  W  T  F  S        S  M  T  W  T  F  S
                  1                 1  2  3  4  5
2  3  4  5  6  7  8        6  7  8  9 10 11 12
9 10 11 12 13 14 15       13 14 15 16 17 18 19
16 17 18 19 20 21 22      20 21 22 23 24 25 26
23 24 25 26 27 28 29      27 28 29 30
30 31

         MAY                       JUNE
S  M  T  W  T  F  S        S  M  T  W  T  F  S
            1  2  3        1  2  3  4  5  6  7
4  5  6  7  8  9 10        8  9 10 11 12 13 14
11 12 13 14 15 16 17      15 16 17 18 19 20 21
18 19 20 21 22 23 24      22 23 24 25 26 27 28
25 26 27 28 29 30 31      29 30

         JULY                     AUGUST
S  M  T  W  T  F  S        S  M  T  W  T  F  S
      1  2  3  4  5                       1  2
6  7  8  9 10 11 12        3  4  5  6  7  8  9
13 14 15 16 17 18 19      10 11 12 13 14 15 16
20 21 22 23 24 25 26      17 18 19 20 21 22 23
27 28 29 30 31            24 25 26 27 28 29 30
                          31

      SEPTEMBER                  OCTOBER
S  M  T  W  T  F  S        S  M  T  W  T  F  S
   1  2  3  4  5  6                 1  2  3  4
7  8  9 10 11 12 13        5  6  7  8  9 10 11
14 15 16 17 18 19 20      12 13 14 15 16 17 18
21 22 23 24 25 26 27      19 20 21 22 23 24 25
28 29 30                  26 27 28 29 30 31

       NOVEMBER                  DECEMBER
S  M  T  W  T  F  S        S  M  T  W  T  F  S
                  1           1  2  3  4  5  6
2  3  4  5  6  7  8        7  8  9 10 11 12 13
9 10 11 12 13 14 15       14 15 16 17 18 19 20
16 17 18 19 20 21 22      21 22 23 24 25 26 27
23 24 25 26 27 28 29      28 29 30 31
30
```

1959

```
       JANUARY                    FEBRUARY
S  M  T  W  T  F  S        S  M  T  W  T  F  S
            1  2  3        1  2  3  4  5  6  7
4  5  6  7  8  9 10        8  9 10 11 12 13 14
11 12 13 14 15 16 17      15 16 17 18 19 20 21
18 19 20 21 22 23 24      22 23 24 25 26 27 28
25 26 27 28 29 30 31

        MARCH                      APRIL
S  M  T  W  T  F  S        S  M  T  W  T  F  S
1  2  3  4  5  6  7                    1  2  3  4
8  9 10 11 12 13 14        5  6  7  8  9 10 11
15 16 17 18 19 20 21      12 13 14 15 16 17 18
22 23 24 25 26 27 28      19 20 21 22 23 24 25
29 30 31                  26 27 28 29 30

         MAY                       JUNE
S  M  T  W  T  F  S        S  M  T  W  T  F  S
               1  2           1  2  3  4  5  6
3  4  5  6  7  8  9        7  8  9 10 11 12 13
10 11 12 13 14 15 16      14 15 16 17 18 19 20
17 18 19 20 21 22 23      21 22 23 24 25 26 27
24 25 26 27 28 29 30      28 29 30
31

         JULY                     AUGUST
S  M  T  W  T  F  S        S  M  T  W  T  F  S
         1  2  3  4                             1
5  6  7  8  9 10 11        2  3  4  5  6  7  8
12 13 14 15 16 17 18       9 10 11 12 13 14 15
19 20 21 22 23 24 25      16 17 18 19 20 21 22
26 27 28 29 30 31         23 24 25 26 27 28 29
                          30 31

      SEPTEMBER                  OCTOBER
S  M  T  W  T  F  S        S  M  T  W  T  F  S
      1  2  3  4  5                       1  2  3
6  7  8  9 10 11 12        4  5  6  7  8  9 10
13 14 15 16 17 18 19      11 12 13 14 15 16 17
20 21 22 23 24 25 26      18 19 20 21 22 23 24
27 28 29 30               25 26 27 28 29 30 31

       NOVEMBER                  DECEMBER
S  M  T  W  T  F  S        S  M  T  W  T  F  S
1  2  3  4  5  6  7                 1  2  3  4  5
8  9 10 11 12 13 14        6  7  8  9 10 11 12
15 16 17 18 19 20 21      13 14 15 16 17 18 19
22 23 24 25 26 27 28      20 21 22 23 24 25 26
29 30                     27 28 29 30 31
```

1959 Vehicle Identification Number (VIN) J59S100001 to J59S109670, total vehicles 9,670. Plate attaches with ordinary Phillips screws to driver's side door jamb, just below upper hinge.
J=Corvette
59=model year
S=St. Louis
Six digit sequential serial number starting at 100001

AIR CLEANERS

Both the two four barrel and the single four barrel used a similar louvered air cleaner approximately 14 inches in diameter. There were three rows of louvers in the lightly polished aluminum top. The top was crimped to the base, therefore neither unit had a serviceable filter. The tops were held in place with wing nuts at the carburetor centers. The wing nuts sat in depressions in the top, and the two four barrel unit had an additional depression in the center.

The fuel injection air cleaner is a two piece unit held together with a wing nut. This provided access to the serviceable filter element. The units are all metal and have a large X stamped in the lid that extends out to the very edge of the air cleaner. The units did not bolt directly to the engine, but to the driver's side inner fender.

AIR CONDITIONING

Not Available

ANTENNAS

Assembly part # 3749220. Three section mast with a plastic cap containing raised ridges at the base, a washer and nut (not visible installed), a plastic spacer and gasket. Mast will have a ball at the top and may or may not have a ring at the top of the lower sections.

BALANCERS

Narrow, 6" diameter balancer with the pulley bolted directly to it.

BATTERIES

1980458 tar top. Year/month and factory of origin is stamped in the top. Cover above the tar on top should be 3 1/2 to 4 inches wide. Caps are all yellow with Delco on the outside early then changed to yellow with Delco in black, visible casting line/flashing is on the bottom, inside. Case side should say "Delco Original Equipment Line" and should be unpainted (warranty replacements were painted). The characters 2SMR53 should be stamped in the side and CAT NO 458 on top of the battery.

BATTERY CABLES

Negative cable is an uninsulated woven strap with an N on the bolt style terminal connection. Positive cable is black vinyl covered with a P on the bolt style terminal connection.

BELLHOUSINGS

3733365, cast iron and date coded.

BRAKE MASTER CYLINDERS

Undated. Part # 5454480 located in the metal plug in the end of the cylinder. Casting # 5456022 in equal size characters located on the driver's side just above the 1" bore size identification. The characters "022" were larger in size late in the year. Filler cap was metal, not plastic, and contains USE GMC BRAKE FLUID around the top.

CARBURETORS

Model numbers are embossed on a brass tag attached under one of the carburetor top screws. The date is also shown on this tag in the form month/year, using A=Jan, B=Feb, etc., an additional digit indicated factory production information.

All 230 HP used WCFB 2818S. The 2818S has the choke mounted at the bottom, and the choke line is attached to the center of the choke cover. A secondary counterweight is used. The main body is cast 0-208, the top 6-1396 or 6-1208.

All 245 HP used WCFB 2626S front, WCFB 2627S rear. All 270 HP used WCFB 2613S front and 2614S rear. All contain secondary counterweights, but no idle air screws. A choke was mounted on the top of the rear carburetor only, and the line was attached at the side, not on the cover. Main bodies are cast 0-049, tops 6-1299 or 0-1049.

CARPET

Extremely short, tight loop, Daytona weave with foam backing. Sectionally sewn to fit the individual areas and contours of the interior. Has an almost mat-like appearance.

COILS

1115091 for carbureted cars, with 091 in raised characters on the case. 1115107 with fuel injection, 107 in raised characters on the case.

CONVERTIBLE TOP

All tops had the rear window sewn into place. The rear window contained the Vinylite name and trademark, the AS 6 identification, cleaning instructions, and a date code hot stamped into the outside surface. The rear bow should match the body contour of the top compartment lid, and requires the use of filler weatherstrips that are somewhat shorter than the 1961-1962 type. The straps attach to the rear bow with a tack strip, not a clamp. The header should be the 59-62 style with a single angle bracket attachment to the side frame. Front hold down latches should have a short handle (1 7/8" across top), not the later 2 1/2" replacement handle. The base of the latch is rounded compared to the tapered angle of the hardtop latch.

CRANKCASE VENTS

The system consists of the oil filler cap and the road draft tube. The road draft tube between the top and the spark plug grommet holder was primarily straight except where it actually attaches to the motor, the remaining lower area had several slight bends and appears almost curved. See the separate oil cap listing.

DISTRIBUTORS

The model used with the 230HP base engine contains a metal tag that wraps around the neck of the distributor below the cap. This contains the date and model number 1110946. All 245 and 270HP cars used model number 1110891 that, along with the date, was stamped on a black and aluminum tag riveted to the distributor housing. The fuel injection cars used models with a cross shaft access cover. The date and model number were stamped in the cover. The 250HP used 1110915. The 290HP used 1110914. Fuel injection distributors all had a facility for driving the tach cable. All models were dated year/month/day.

EMBLEMS

Nose & Trunk (Gold Letters)

Side

Side

ENGINE BLOCKS

Casting numbers 3737739 for early cars, or 3756519. All engines were 283 cubic inches. Casting number is located on the driver's side rear flange where the bellhousing attaches. The date is located directly opposite on the passenger side rear flange. They are dated month/day/year, with a single digit year representation. Double digits indicate Tonawanda manufacture and are not correct for Corvette. The month is indicated by a consecutive alphabetic character, ie January = A, February = B, etc. A code indicating the assembly location, date, and engine characteristics was stamped on a pad located on the passenger side of the motor. It is just below the cylinder head, next to the water pump. The first character should be F for the Flint assembly plant. The numeric characters represent the date, and the last two characters indicate the engine type as shown in the chart that follows.

In 1959 the pad did not contain a serial number.
CQ = STD 230HP, 4 barrel, manual transmission
DG = STD 230HP, 4 barrel, automatic transmission
CT = 469 245HP, 2 four barrels, manual transmission
DJ = 469 245HP, 2 four barrels, automatic transmission
CR = 579 250HP, fuel injection, manual transmission
DH = 579 250HP, fuel injection, automatic transmission
CU = 469 270HP, two four barrels, special cam, manual
CS = 579 290HP, fuel injection, special cam, manual

EXHAUST MANIFOLDS

Dated month and day only for correct Flint manifolds. Casting # 3749965 left, 3750556 right.

FUEL FILTERS

All cars with two four barrels used a glass sediment bowl style of filter. A 'stone' filter element was used and the bowl was held in place by a wire bail. The primarily flat cast alloy top contained a small AC logo. Cars with one four barrel used a filter mounted in the carburetor just behind the inlet fitting.

Fuel injected cars used a screw together, canister style filter with a removable element. Both fuel lines connected to the sides of the top portion of the assembly, and the unit was mounted behind the fuel meter.

FUEL INJECTION

Six different units were used in 1959. All units had a ribbed plenum that contained a black/aluminum tag showing the part number and a serial number on the driver's side. The serial number has no relationship to any other number. A snowflake was not cast into the plenum top. Cold start enrichment was assisted by an electric thermostatic coil mounted on the air meter. See unit description for additional details. A conventional choke with it's associated butterfly was not used.

7014900   250HP. Cranking signal valve mounted in the plenum was part of the cold enrichment. Balance tube ran from the fuel meter to the air cleaner adaptor. Two lines ran from the main diaphragm, one to the air meter, and one to the cranking signal valve.
Air meter stamped 7014901, cast 7014922
Fuel meter stamped 7014902, cast 7014889

7014900R   290HP. Merely a recalibrated 7014900 unit.
Air meter stamped 7014901, cast 7014922
Fuel meter stamped 7014902, cast 7014889

7017200   250HP. Cranking signal valve mounted in the plenum was part of the cold enrichment. Balance tube ran from the fuel meter to the air cleaner adaptor. Two lines ran from the main diaphragm, one to the air meter,

	and one to the cranking signal valve. L, M, and possible N nozzle usage. Air meter stamped 7017201, cast 7014922 Fuel meter stamped 7017202, cast 7014889
7017250	290HP. Cranking signal valve mounted in the plenum was part of the cold enrichment. Balance tube ran from the fuel meter to the air cleaner adaptor. Two lines ran from the main diaphragm, one to the air meter, and one to the cranking signal valve. Air meter stamped 7017251, cast 7014922 Fuel meter stamped 7017252, cast 7014889
7017300	290HP. Cold start enrichment was provided by a micro switch activated solenoid, therefore no cranking signal valve was used. Forward balance tube ran from the fuel meter to the air cleaner adaptor. One line ran from the main diaphragm to the air meter. Air meter stamped 7017301, cast 7014922 Fuel meter stamp 7017302, 7014802, 7014965, cast 7014312
7017300R	250HP. A recalibrated 7017300 unit. Air meter stamped 7017301, cast 7014922 Fuel meter stamp 7017302, 7014802, 7014965, cast 7014312

FUEL PUMPS

The AC logo is cast directly in the top of the pump. The four digit Delco number is stamped in the edge of the mounting flange and should be 4656. Pumps are single diaphragm, screwed together assemblies, using short, 3/4" screws.

GAUGES

Fuel/Temp - Battery/Oil combinations on the small gauges contain concave lenses with line indicators on a metal face. Pointers are rounded and a flat chrome gauge name plate covers the lower half of the gauge face. All tachs are 7000 RPM with white characters on a black metal face with yellow and orange warning areas. Clocks have a square back and are made by Westclox (stamped on the back). The F for the fast/slow indicator should be on the left. Speedometers are 160 MPH having a metal inner face with lines, and a flat outer face with numbers.

GENERATORS

1102043/30 Amp, all except 290HP
1102059/30 Amp, early 290HP
1102173/35 Amp, late with 290HP

The model number and a date code appear on a red/aluminum tag attached to the generator. The year, month, day, are listed as serial number on the tag, and the month appears as an alphabetic character with A=Jan, B=Feb, etc. All units use a one piece fan-pulley assembly. All 1102043 applications used a 3 5/8" pulley with a 1/2" belt groove and a spacer between the fan and pulley. The 1102059 and 1102173 used a 4" pulley with a 1/2" belt groove. Tach drive assembly is attached to the rear on all but 290HP models.

GLASS

All laminated glass was LOF Safety Plate and, along with the logo, was designated as such by AS1 for windshields and AS2 for side glass. Side and rear Plexiglas was AS4. A two character code (month-year) also appeared indicating the date of manufacture. January=N, February=X, March=L, April=G, May=J, June=I, July=U, August=T, September=A, October=Y, November=C, December=V. The letter N indicates 1958, the letter Y 1959. Tinted glass was not used.

GRILLE

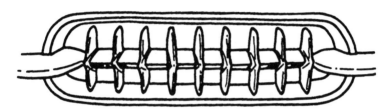

HARDTOP

Should have a 2" wide, two piece, stainless header trim strip with a small cap covering the center joint. The lower rear stainless trim consists of two pieces joined at the center of the rear window by a cap. This cap, and lower trim will be straight across to match the top compartment lid. All AS-4 Plexiglas windows are dated month/year. Front hold down latches should have a short handle (1 7/8" across top), not the later 2 1/2" replacement handle. The base of the latch is a tapered angle compared to the rounded soft top latch. Headliner material should be the same color and grain pattern as the seats, etc.

HEADS

Casting # 3755550 for all except very late production, late production used 3767465. Dated month/day/year, with a single digit year representation. Double digits indicate Tonawanda manufacture and are not correct for Corvette. The month is indicated by a consecutive alphabetic character, ie January = A, February = B, etc. Symbol at the end of the head is a partial indication of the application.

HORNS

The last three digits of the part number are stamped into the horn. The high note is 9000352, the low note 9000351. The production date is not stamped. Mounting brackets were welded to the horn, not bolted.

HORN RELAYS

1116781 for all cars. The characters '781', for the last three digits of the part number, and '12VN', for 12 volt, negative ground, are stamped in the mounting flange. The cover is stamped with Delco Remy.

HUBCAPS

Standard

RPO

IGNITION SHIELDING

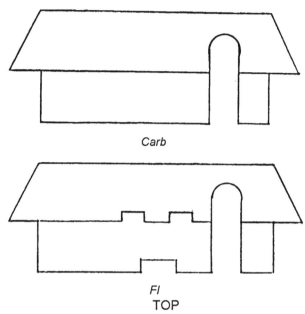

Carb

FI
TOP

All shields were stainless steel and had a front skirt that was visibly shorter than the rear. There should be a depression in the upper rear skirt near the middle. All rear attachments are through holes, not slots. For carbureted cars the front and rear skirt are straight across, and only contain an opening for the accelerator linkage. For cars with fuel injection there is a bulge in the front skirt to accommodate the coil, a single notch in the rear skirt, and two notches in the front skirt.

VERTICALS

All left shields should have the shorter, straight bottom. Both sides are made from chrome plated steel. The right shield lower mounting should be through a hole.

HORIZONTALS

A single piece of shielding was used on each side. Both sides have the smaller exhaust manifold cut-outs, and no reinforcement bar.

INTAKE MANIFOLDS

Dated month/day/year. Casting numbers are located on the top surface. Cast iron manifold date codes are also located on the top surface, and are therefore visible with the manifold installed. Correct manifolds should have a single digit year representation. Aluminum manifold date codes are cast into the bottom surface and are not visible unless the manifold is removed from the engine. Cast iron 230HP manifold was 3746829. The aluminum manifold was 3739653 for both the 245 and 270HP applications.

JACK

Correct scissor jack has an identification number, SJ4653, cast directly into the top of the cast iron base. Jack arms and saddle are stamped steel. Jack screw has fine threads and the handle facility is an oblong slot in the end.

KNOBS

LAMPS

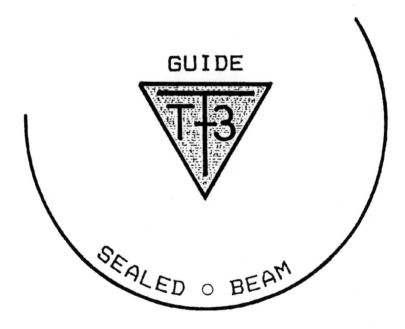

MIRRORS

Outside rearview mirror was stamped 'GUIDE Y-50' on the mirror back at the ball stud. This ball should be centered in the back of the mirror. The glass was not dated, and considered non replaceable since the later style snap ring access was not provided.

Inside mirrors had a base containing a threaded top where the mirror was attached and held in place with a locknut. The mirror was an exact rectangular shape with rounded corners and three rivets surrounding the ball stud in back. A 1/8" circle with an S in it appears 3/16" above the center rivet. The mirror glass was not dated.

OIL CAPS

Solid lifter engines were non vented

OPTIONS

		Quantity	Price
2934	Base convertible	9,670	$3,875.00
Incl.	283, 230HP	5,487	NC
Incl.	3 Speed	3,617	NC
Incl.	Black folding top	2,190	NC
101	Heater	8,909	102.25
102	Wonderbar AM radio	7,001	149.80
107	Parking brake alarm	3,601	5.40
108	Courtesy light	3,601	6.50
109	Windshield washers	7,929	16.15
121	Radiator fan clutch	67	21.55
261	Sunvisors	3,722	10.80
276	15X5 1/2 Wheels	214	NC
290	Whitewall tires	8,173	31.55
313	Powerglide	1,878	199.10
419	Hardtop	3,786	236.75
419	Hardtop without soft top	1,695	NC
426	Power windows	587	59.20
440	Two tone paint	2,931	16.15
469	283, 245HP engine	1,417	150.65
469C	283, 270HP engine	1,846	182.95
470	White soft top	4,092	NC
470	Blue soft top	412	NC
470	Turquoise soft top	217	NC
473	Power folding top	661	139.90
579	283, 250HP engine	172	484.20

579D	283, 290HP engine	920	484.20
675	Positraction rear	4,170	48.45
684	HD Brakes & suspension	142	425.05
685	4 Speed transmission	4,175	188.30
686	Metallic brakes	333	26.90
1408	Blackwall nylon tires		15.75

DEALER AVAILABLE OPTIONS
Tool Kit
Spotlight

PAINT/TRIM

1956-1962 Corvettes did not have a tag that provided paint and trim code information.

EXTERIOR	COVE	INTERIOR	SOFT TOP
Tuxedo Black	Silver	Red:Black:Blue	Black:White
Roman Red	White	Red:Black	White:Black
Classic Cream	White	Black	Black:White
Snowcrest White	Silver	Red:Blue:Black:Turquoise	Black:White:Blue:Turquoise
Frost Blue	White	Blue:Red	Blue:Black:White
Inca Silver	White	Red:Black	Black:White
Crown Sapphire	White	Turquoise	Black:White:Turquoise

Cove colors shown were optional, body color standard.
These were factory recommended combinations.

PRODUCTION FIGURES

	MONTHLY	CUMULATIVE
September	409	409
October	223	632
November	955	1587
December	1054	2641
January	1321	3962
February	959	4921
March	1112	6033
April	1111	7144
May	790	7934
June	768	8702
July	735	9437
August	233	9670

RADIATORS

All radiators were copper and contained an identification tag soldered to the top tank. This tag contained the part number 3141674, and the date of manufacture represented by a double digit year code and a single alphabetic month code (Jan=A, Feb=B, etc.). The fan shroud was the wide, four piece style.

RADIATOR CAPS

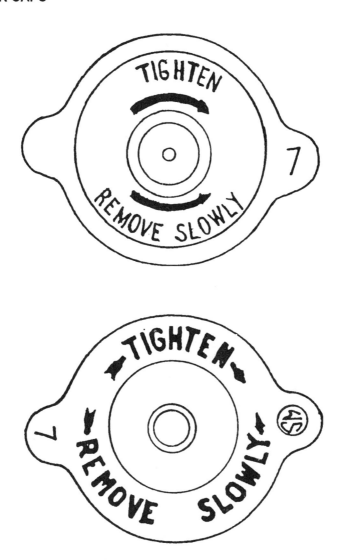

RADIOS

Model number 987730, 7 tube Wonder Bar. The face of the push buttons should be flat. A paper tag on the outside of one of the radio covers contains the model number.

REAR AXLE

Code and build date (month/day) are located on the right, front of the carrier.
CODES
AE .... 3.55, Open, Powerglide
AH .... 3.70, Open, manual
AN .... 3.70, Positraction, manual
AP .... 4.11, Positraction, manual
AQ .... 4.56, Positraction, manual
AS .... 3.70, Positraction, manual, HD brakes & suspension
AT .... 4.11, Positraction, manual, HD brakes & suspension
AU .... 4.56, Positraction, manual, HD brakes & suspension
FJ .... 3.70, Open, manual, metallic brakes
FK .... 3.70, Positraction, manual, metallic brakes
FL .... 4.11, Positraction, manual, metallic brakes
FM .... 4.56, Positraction, manual, metallic brakes

SEATS

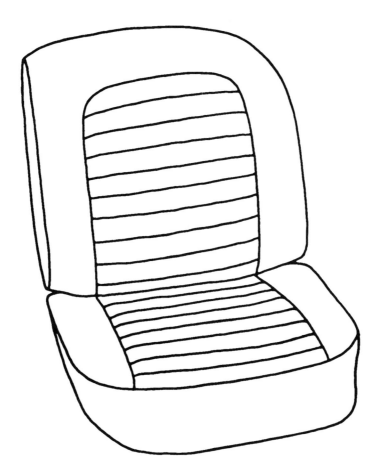

SHIFTERS

The three speed shifter and the automatic transmission shifter have a similar overall appearance even including the mounting bracket. Both utilized a round, thin, chrome plated shift lever. Automatic shift pattern was L D N R P front to back.

The four speed shifter was also somewhat similar even with the addition of the reverse lockout. The shifter tower shaft contained a single spring for gate control tension. This was held in place by a round, slotted nut. An anti rattle spring was attached between the shift lever and a small bracket bolted to the bottom of the shifter support. The reverse lever should be distinctly longer than the third/fourth lever, unlike the 1963 shifters where they are about the same length.

SPARK PLUG WIRES

Distributor boot is 180 degree, black. Spark plug boot is 90 degree, black. Wire is 7MM black with the imprint RADIO GM TVRS, undated.

STARTERS

1107664 was used on all cars. The date and model number were stamped directly into the starter housing. The date is shown as year, month, day, and the month appears as an alphabetic character with A=Jan, B=Feb, etc. The solenoid bakelite should contain Delco Remy.

STEERING WHEELS

Color matching (red, black, turquoise or blue) wheels were plastic with metal spokes. Outer plastic circumference should be free of grind marks found on reproduction wheels. Finishing marks on the spokes should run the same direction as the rim on the face of the spoke and perpendicular to the rim on the back. The amount of simulated grain, or lack of it, appearing in the plastic of original wheels is a point of controversy. Replacement GM wheels have plenty, but some very original cars have little or none. The idea that the grain has worn away is unlikely since it is typically missing from all areas of the wheel, even areas with little hand contact.

THERMOSTAT HOUSINGS

Engines with a cast iron intake manifold used a curved, cast iron housing, number 3711268. Engines with an aluminum intake manifold used an angled, aluminum housing, number 3837223. The aluminum housing without metal saver notches around the stud bosses is clearly correct for this application. These housings appear in factory photos and on original cars through 1962. Chevrolet provided a slightly different part with the same number as a replacement. Metal saver notches are around the stud bosses and the manufacturers logo appears along with the part number. These may or may not have been used originally.

TIRES

Firestone 'Deluxe Champion', US Royal 'Air Ride', or B.F. Goodrich 'Silvertown' 6.70x15, tubeless. If whitewall equipped, the width should be 2" to 2 9/16" nominal. DOT (department of transportation) information did not appear in the sidewall of original tires.

TRANSMISSIONS

The Borg Warner four speed was available as an option. The aluminum tail housings should not have two half moon cuts at the rear bushing. There should be only one shifter mounting position (3 holes), and M should be cast along with the date. The side cover is dated inside and should contain webbing. The main case is cast iron and has T10-1 and month/day cast into the passenger side until May when T10-1B started. A build code is stamped on the upper rear, driver's side, unmachined part of the case. The code is in the form: location/month/day/year/shift, W=Warner Gear, month=Jan=A, Feb=B, etc., shift=1,2,3. Vehicle serial numbers are not stamped on these transmissions.

Build information is stamped on the upper rear of the case, passenger side for three speeds, and on the rear flange of the governor cover for Powerglide. Build information is in the form: location/month/day/shift (D or N). S=Saginaw (three speed), C=Cleveland (Powerglide). The three speed must have a shifter mounting position on the tail housing.

VALVE COVERS

Until May all valve cover mounting holes were closer together on top than they were on the bottom. After mid May they were equally spaced at 8 3/4" top and bottom. The 230HP engine used a stamped steel cover with CHEVROLET in raised script.

The remaining engines used a cast aluminum cover with CORVETTE in raised, 1/8" script. A casting flaw should not appear through the 'O' in CORVETTE. All covers had seven fins running the length of the cover. Two notches appeared on the inside lip for intake manifold clearance.

VOLTAGE REGULATORS

1119001, or 1119002 if equipped with 1102173 generator. The cover was stamped with Delco Remy and was attached with slotted cap screws. The part number and date, along with '12VN' for 12 volt negative ground, are stamped in the mounting flange. The date is shown as year/month, with the month represented by an alphabetic character, ie A=Jan, B=Feb, etc.

WATER PUMPS

Casting # 3736493, undated.

WIPER ARMS & BLADES

All cars used a bright stainless, Trico brand, with rubber insert. The Trico name may not always appear on the parts. The rubber inserts have a series of raised dots running the length of both sides. The blade construction consists of a main bridge piece that is somewhat shorter than the overall length of the blade. Attached to this are three links that hold the rubber insert. There are also two hard rubber blocks at each end of the blade.

WIPER MOTORS

5044266 stamped in the front cover. There should not be a ground wire riveted to the rear cover arm. There should be a 1" long metal support stand attached to the lower armature housing bolt. Windshield washers were optional equipment.

CORVETTE
'60

1959

JANUARY
```
S  M  T  W  T  F  S
            1  2  3
 4  5  6  7  8  9 10
11 12 13 14 15 16 17
18 19 20 21 22 23 24
25 26 27 28 29 30 31
```

FEBRUARY
```
S  M  T  W  T  F  S
 1  2  3  4  5  6  7
 8  9 10 11 12 13 14
15 16 17 18 19 20 21
22 23 24 25 26 27 28
```

MARCH
```
S  M  T  W  T  F  S
 1  2  3  4  5  6  7
 8  9 10 11 12 13 14
15 16 17 18 19 20 21
22 23 24 25 26 27 28
29 30 31
```

APRIL
```
S  M  T  W  T  F  S
          1  2  3  4
 5  6  7  8  9 10 11
12 13 14 15 16 17 18
19 20 21 22 23 24 25
26 27 28 29 30
```

MAY
```
S  M  T  W  T  F  S
                1  2
 3  4  5  6  7  8  9
10 11 12 13 14 15 16
17 18 19 20 21 22 23
24 25 26 27 28 29 30
31
```

JUNE
```
S  M  T  W  T  F  S
    1  2  3  4  5  6
 7  8  9 10 11 12 13
14 15 16 17 18 19 20
21 22 23 24 25 26 27
28 29 30
```

JULY
```
S  M  T  W  T  F  S
          1  2  3  4
 5  6  7  8  9 10 11
12 13 14 15 16 17 18
19 20 21 22 23 24 25
26 27 28 29 30 31
```

AUGUST
```
S  M  T  W  T  F  S
                   1
 2  3  4  5  6  7  8
 9 10 11 12 13 14 15
16 17 18 19 20 21 22
23 24 25 26 27 28 29
30 31
```

SEPTEMBER
```
S  M  T  W  T  F  S
       1  2  3  4  5
 6  7  8  9 10 11 12
13 14 15 16 17 18 19
20 21 22 23 24 25 26
27 28 29 30
```

OCTOBER
```
S  M  T  W  T  F  S
                1  2  3
 4  5  6  7  8  9 10
11 12 13 14 15 16 17
18 19 20 21 22 23 24
25 26 27 28 29 30 31
```

NOVEMBER
```
S  M  T  W  T  F  S
 1  2  3  4  5  6  7
 8  9 10 11 12 13 14
15 16 17 18 19 20 21
22 23 24 25 26 27 28
29 30
```

DECEMBER
```
S  M  T  W  T  F  S
       1  2  3  4  5
 6  7  8  9 10 11 12
13 14 15 16 17 18 19
20 21 22 23 24 25 26
27 28 29 30 31
```

1960

JANUARY
```
S  M  T  W  T  F  S
                1  2
 3  4  5  6  7  8  9
10 11 12 13 14 15 16
17 18 19 20 21 22 23
24 25 26 27 28 29 30
31
```

FEBRUARY
```
S  M  T  W  T  F  S
    1  2  3  4  5  6
 7  8  9 10 11 12 13
14 15 16 17 18 19 20
21 22 23 24 25 26 27
28 29
```

MARCH
```
S  M  T  W  T  F  S
       1  2  3  4  5
 6  7  8  9 10 11 12
13 14 15 16 17 18 19
20 21 22 23 24 25 26
27 28 29 30 31
```

APRIL
```
S  M  T  W  T  F  S
                1  2
 3  4  5  6  7  8  9
10 11 12 13 14 15 16
17 18 19 20 21 22 23
24 25 26 27 28 29 30
```

MAY
```
S  M  T  W  T  F  S
 1  2  3  4  5  6  7
 8  9 10 11 12 13 14
15 16 17 18 19 20 21
22 23 24 25 26 27 28
29 30 31
```

JUNE
```
S  M  T  W  T  F  S
          1  2  3  4
 5  6  7  8  9 10 11
12 13 14 15 16 17 18
19 20 21 22 23 24 25
26 27 28 29 30
```

JULY
```
S  M  T  W  T  F  S
                1  2
 3  4  5  6  7  8  9
10 11 12 13 14 15 16
17 18 19 20 21 22 23
24 25 26 27 28 29 30
31
```

AUGUST
```
S  M  T  W  T  F  S
    1  2  3  4  5  6
 7  8  9 10 11 12 13
14 15 16 17 18 19 20
21 22 23 24 25 26 27
28 29 30 31
```

SEPTEMBER
```
S  M  T  W  T  F  S
             1  2  3
 4  5  6  7  8  9 10
11 12 13 14 15 16 17
18 19 20 21 22 23 24
25 26 27 28 29 30
```

OCTOBER
```
S  M  T  W  T  F  S
                   1
 2  3  4  5  6  7  8
 9 10 11 12 13 14 15
16 17 18 19 20 21 22
23 24 25 26 27 28 29
30 31
```

NOVEMBER
```
S  M  T  W  T  F  S
       1  2  3  4  5
 6  7  8  9 10 11 12
13 14 15 16 17 18 19
20 21 22 23 24 25 26
27 28 29 30
```

DECEMBER
```
S  M  T  W  T  F  S
                1  2  3
 4  5  6  7  8  9 10
11 12 13 14 15 16 17
18 19 20 21 22 23 24
25 26 27 28 29 30 31
```

1960 Vehicle Identification Number (VIN) 00867S100001 to 00867S110261, total vehicles 10,261. Early plates attach with ordinary Phillips screws to driver's side door jamb, just below upper hinge. Later plates are located under the hood, spot welded to the steering column.
First digit=model year
S=St. Louis
Six digit sequential serial number starting at 100001

AIR CLEANERS

Both the two four barrel and the single four barrel used a similar louvered air cleaner approximately 14 inches in diameter. There were three rows of louvers in the lightly polished aluminum top. With early cars the top was crimped to the base, therefore neither unit had a serviceable filter. The balance of the cars had two piece units allowing service of the foam air filter element. The tops were held in place with wing nuts at the carburetor centers. The wing nuts sat in depressions in the top, and the two four barrel unit had an additional depression in the center. Beware of similar, non aluminum, two piece Pontiac air cleaners.

The fuel injection air cleaner is a two piece unit held together with a wing nut. This provided access to the serviceable filter element. The units are all metal and have a large X stamped in the lid that extends out to the very edge of the air cleaner. The units did not bolt directly to the engine, but to the driver's side inner fender.

AIR CONDITIONING

Not Available

ANTENNAS

Assembly part # 3749220. Three section mast with a plastic cap containing raised ridges at the base, a washer and nut (not visible installed), a plastic spacer and gasket. Mast will have a ball at the top and may or may not have a ring at the top of the lower sections.

BALANCERS

Narrow, 6" diameter balancer with the pulley bolted directly to it.

BATTERIES

1980458 tar top. Year/month and factory of origin is stamped in the top. Cover above the tar on top should be 3 1/2 to 4 inches wide. Caps are all yellow with Delco in black on the outside, visible casting line/flashing on the bottom, inside. Case side should say "Delco Original Equipment Line" and should be unpainted (warranty replacements were painted). The characters 2SMR53 should be stamped in the side and CAT NO 458 on top of the battery.

BATTERY CABLES

Negative cable is an uninsulated woven strap with an N on the bolt style terminal connection. Positive cable is black vinyl covered with a P on the bolt style terminal connection.

BELLHOUSINGS

3764591, aluminum without date code.

BRAKE MASTER CYLINDERS

Undated. Part # 5454480 located in the metal plug in the end of the cylinder. Casting # 5456022 located on the driver's side just above the 1" bore size identification. The characters "022" were larger than the others. Filler cap was metal, not plastic, and contains USE GMC BRAKE FLUID around the top.

CARBURETORS

Model numbers are embossed on a brass tag until October, after that an aluminum tag was used. They were attached under one of the carburetor top screws. The date is also shown on this tag in the form month/year, using A=Jan, B=Feb, etc., an additional digit indicated factory production information.

All 230 HP used WCFB 2818S early, 3059S for the balance of the year. Both have the choke mounted at the bottom, and the choke line is attached to the center of the choke cover. A secondary counterweight is used. The main body is cast 0-208, the top 6-1396 on the 2818S, and 0-1343 or 0-1517 main body, 6-1396 top on the 3059S.

All 245 HP used WCFB 2626S front, WCFB 2627S rear. All 270 HP used WCFB 2613S front and 2614S rear. All contain secondary counterweights, but no idle air screws. A choke was mounted on the top of the rear carburetor only, and the line was attached at the side, not on the cover. Main bodies are cast 0-049, tops 6-1299 or 0-1049.

CARPET

Salt and pepper appearing loop carpet, foam backed, and sectionally sewn to fit the individual sections and contours of the interior.

COILS

1115091 for carbureted cars, with 091 in raised characters on the case. 1115107 with fuel injection, 107 in raised characters on the case.

CONVERTIBLE TOP

All tops had a rear window attached with heat sealing from the inside. The rear window contained the Vinylite name and trademark, the AS 6 identification, cleaning instructions, and a date code hot stamped into the outside surface. The rear bow should match the body contour of the top compartment lid, and requires the use of the filler weatherstrips that are somewhat shorter than the 1961-1962 type. The straps attach to the rear bow with a tack strip, not a clamp. The header should be the 59-62 style with a single angle bracket attachment to the side frame. Front hold down latches should have a short handle (1 7/8" across top), not the later 2 1/2" replacement handle. The base of the latch is rounded compared to the tapered angle of the hardtop latch.

CRANKCASE VENTS

The system consists of the oil filler cap and the road draft tube. The road draft tube between the top and the spark plug grommet holder was primarily straight except where it actually attaches to the motor, the remaining lower area had several slight bends and appears almost curved. See the separate oil cap listing.

DISTRIBUTORS

The model used with the 230HP base engine contains a metal tag that wraps around the neck of the distributor below the cap. This contains the date and model number 1110946. All 245 and 270HP cars used model number 1110891 that, along with the date, was stamped on a black and aluminum tag riveted to the distributor housing. The fuel injection cars used models with a cross shaft access cover. The date and model number were stamped in the cover. The 250HP used 1110915. The 290HP used 1110914. Fuel injection distributors all had a facility for driving the tach cable. All models were dated year/month/day.

EMBLEMS

Nose & Trunk (Gold Letters)

Side

Side

ENGINE BLOCKS

Casting # 3756519. All engines were 283 cubic inches. Casting number is located on the driver's side rear flange where the bellhousing attaches. The date is located directly opposite on the passenger side rear flange. They are dated month/day/year, with a single digit year representation. Double digits indicate Tonawanda manufacture and are not correct for Corvette. The month is indicated by a consecutive alphabetic character, ie January = A, February = B, etc. A code indicating the assembly location, date, and engine characteristics was stamped on a pad located on the passenger side of the motor. It is just below the cylinder head, next to the water pump. The first character should be F for the Flint assembly plant. The numeric characters represent the date, and the last two characters indicate the engine type as shown in the chart that follows. Starting about mid production in 1960 the pad did contain a vehicle serial number, separate from, and to the left of the assembly code.

CQ = STD 230HP, 4 barrel, manual transmission
DG = STD 230HP, 4 barrel, automatic transmission
CT = 469 245HP, 2 four barrels, manual transmission
DJ = 469 245HP, 2 four barrels, automatic transmission
CR = 579 250HP, fuel injection, manual transmission
CU = 469 270HP, 2 four barrels, special cam, manual
CS = 579 290HP, fuel injection, special cam, manual

EXHAUST MANIFOLDS

Dated month and day only for correct Flint manifolds. Casting # 3749965 left, 3750556 right.

FUEL FILTERS

All cars with two four barrels used a glass sediment bowl style of filter. A 'stone' filter element was used and the bowl was held in place by a wire bail. The primarily flat cast alloy top contained a small AC logo. Cars with one four barrel used a filter mounted in the carburetor just behind the inlet fitting.

Fuel injected cars used a screw together, canister style filter with a removable element. Both fuel lines connected to the sides of the top portion of the assembly, and the unit was mounted behind the fuel meter.

FUEL INJECTION

Five different units were used in 1960. All units had a ribbed plenum except 7017310 and 7017320 which were flat. They all contained a black/aluminum tag showing the part number and a serial number on the driver's side. The serial number has no relationship to any other number.

A snowflake was not cast into the plenum top. Cold start enrichment was assisted by an electric thermostatic coil mounted on the air meter. See unit description for additional details. A conventional choke with it's associated butterfly was not used.

7017200   250HP. Cranking signal valve mounted in the plenum was part of the cold enrichment. Balance tube ran from the fuel meter to the air cleaner adaptor. Two lines ran from the main diaphragm, one to the air meter, and one to the cranking signal valve.
Air meter stamped 7017201, cast 7014922
Fuel meter stamped 7017202, cast 7014889

7017250   290HP. Cranking signal valve mounted in the plenum was part of the cold enrichment. Balance tube ran from the fuel meter to the air cleaner adaptor. Two lines ran from the main diaphragm, one to the air meter, and one to the cranking signal valve.

7017300	Air meter stamped 7017251, cast 7014922 Fuel meter stamped 7017252, cast 7014889 290HP. Cold start enrichment was provided by a micro switch activated solenoid, therefore no cranking signal valve was used. Forward balance tube ran from the fuel meter to the air cleaner adaptor. One line ran from the main diaphragm to the air meter. Usage for 1960 may be questionable.
7017310	Air meter stamped 7017301, cast 7014922 Fuel meter stamp 7017302, 7014802, 7014965, cast 7014312 250HP. Cranking signal valve mounted in the plenum was part of the cold enrichment. Balance tube ran from the fuel meter to the air cleaner adaptor. Two lines ran from the main diaphragm, one to the air meter, and one to the cranking signal valve.
7017320	Air meter stamped 7017201, cast 7014922 Fuel meter stamped 7017202, cast 7014889 290HP. Cranking signal valve mounted in the plenum was part of the cold enrichment. Balance tube ran from the fuel meter to the air cleaner adaptor. Two lines ran from the main diaphragm, one to the air meter, and one to the cranking signal valve. Air meter stamped 7017251, cast 7014922 Fuel meter stamped 7017252, cast 7014889

FUEL PUMPS

The AC logo is cast directly in the top of the pump. The four digit Delco number is stamped in the edge of the mounting flange and should be 4656. Pumps are single diaphragm, screwed together assemblies, using short, 3/4" screws.

GAUGES

Fuel/Temp - Battery/Oil combinations on the small gauges contain concave lenses with line indicators on a metal face. Pointers are pointed and a flat chrome gauge name plate covers the lower half of the gauge face. Temp gauge should be 220 degree. All tachs are 7000 rpm on a metal face containing light green, dark green, yellow, orange, and red (high rpm only) warning areas, and an outer concave lens. Clocks have a square back and are made by Westclox (stamped on the back). The F for the slow/fast indicator should be on the left. Speedometers are 160 MPH having an inner metal face with lines, and a flat outer face with numbers.

GENERATORS

1102043/30 Amp, all except 290HP
1102173/35 Amp, with 290HP

The model number and a date code appear on a red/aluminum tag attached to the generator. The year, month, day, are listed as serial number on the tag, and the month appears as an alphabetic character with A=Jan, B=Feb, etc. All units use a one piece fan-pulley assembly. All 1102043 applications used a 3 5/8" pulley with a 1/2" belt groove and a spacer between the fan and pulley. The 1102173 used a 4" pulley with a 1/2" belt groove. Tach drive assembly is attached to the rear on all but 290HP models.

GLASS

All laminated glass was LOF Safety Plate and, along with the logo, was designated as such by AS1 for windshields and AS2 for side glass. Side and rear Plexiglas was AS4. A two character code (month-year) also appeared indicating the date of manufacture. January=N, February=X, March=L, April=G, May=J, June=I, July=U, August=T, September=A, October=Y, November=C, December=V. The letter Y indicates 1959, the letter U 1960. Tinted glass was not used.

GRILLE

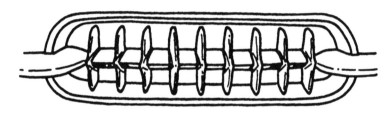

HARDTOP

Should have a 2" wide, two piece, stainless header trim strip with a small cap covering the center joint. The lower rear stainless trim consists of two pieces joined at the center of the rear window by a cap. This cap, and lower trim will be straight across to match the top compartment lid. All AS-4 Plexiglas windows are dated month/year. Front hold down latches should have a short handle (1 7/8" across top), not the later 2 1/2" replacement handle. The base of the latch is a tapered angle compared to the rounded soft top latch. Headliner material should be the same color and grain pattern as the seats, etc.

HEADS

Casting # 3774692. Dated month/day/year, with a single digit year representation. Double digits indicate Tonawanda manufacture and are not correct for Corvette. The month is indicated by a consecutive alphabetic character, ie January = A, February = B, etc. Symbol at the end of the head is a partial indication of the application.

HORNS

The last three digits of the part number are stamped into the horn. The high note is 9000352, the low note 9000351. The production date is not stamped. Mounting brackets were welded to the horn, not bolted.

HORN RELAYS

1116781 for all cars. The characters '781', for the last three digits of the part number, and '12VN', for 12 volt, negative ground, are stamped in the mounting flange. The cover is stamped with Delco Remy.

HUBCAPS

Standard

RPO

IGNITION SHIELDING

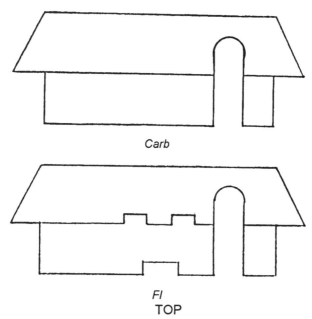

Carb

FI
TOP

All shields were stainless steel and had a front skirt that was visibly shorter than the rear. There should be a depression in the upper rear skirt near the middle. All rear attachments are through holes, not slots. For carbureted cars the front and rear skirt are straight across, and only contain an opening for the accelerator linkage. For cars with fuel injection there is a bulge in the front skirt to accommodate the coil, a single notch in the rear skirt, and two notches in the front skirt.

VERTICALS

All left shields should have the shorter, straight bottom. The left side is stainless, the right is made from chrome plated steel. The right shield lower mounting should be through a hole.

HORIZONTALS

A single piece of shielding is used on each side. Both sides have the smaller exhaust manifold cut-outs, and no reinforcement bar.

INTAKE MANIFOLDS

Dated month/day/year. Casting numbers are located on the top surface. Cast iron manifold date codes are also located on the top surface, and are therefore visible with the manifold installed. Correct manifolds should have a single digit year representation. Aluminum manifold date codes are cast into the bottom surface and are not visible unless the manifold is removed from the engine. Cast iron 230HP manifold was 3746829. The aluminum manifold was 3739653 for both the 245 and 270HP applications.

JACK

Correct scissor jack has an identification number, SJ4653, cast directly into the top of the cast iron base. Jack arms and saddle are stamped steel. Jack screw has fine threads and the handle facility is an oblong slot in the end.

KNOBS

LAMPS

MIRRORS

Outside rearview mirror was stamped 'GUIDE Y-50' on the mirror back at the ball stud. This ball should be centered in the back of the mirror. The glass was not dated, and considered non replaceable since the later style snap ring access was not provided.

Inside mirrors had a base containing a threaded top where the mirror was attached and held in place with a locknut. The mirror was an exact rectangular shape with rounded corners and three rivets surrounding the ball stud in back. A 1/8" circle with an S in it appears 3/16" above the center rivet. The mirror glass was not dated.

OIL CAPS

Solid lifter engines were non vented

OPTIONS

		Quantity	Price
00867	Base convertible	10,261	$3,872.00
Incl.	283, 230HP	5,827	NC
Incl.	3 Speed	3,167	NC
Incl.	Black folding top	4,931	NC
101	Heater	9,808	102.25
102	Wonderbar AM radio	8,166	137.75
107	Parking brake alarm	4,051	5.40
108	Courtesy light	6,774	6.50
109	Windshield washers	7,205	16.15
121	Radiator fan clutch	2,711	21.55
261	Sunvisors	5,276	10.80
276	15X5 1/2 Wheels	246	NC
290	Whitewall tires	9,104	31.55
313	Powerglide	1,766	199.10
419	Hardtop	3,506	236.75
419	Hardtop without soft top	1,641	NC
426	Power windows	544	59.20
440	Two tone paint	3,309	16.15
469	283, 245HP engine	1,211	150.65
469C	283, 270HP engine	2,364	182.95
470	White soft top	4,947	NC
470	Blue soft top	333	NC
473	Power folding top	512	139.90
579	283, 250HP engine	100	484.20
579D	283, 290HP engine	759	484.20

Code	Option	Qty	Price
675	Positraction rear	5,231	43.05
685	4 Speed transmission	5,328	188.30
686	Metallic brakes	920	26.90
687	HD Brakes & steering	119	333.60
1408	Blackwall nylon tires		15.75
1625	24 Gallon fuel tank		161.40

DEALER AVAILABLE OPTIONS
Tool Kit
Spotlight

PAINT/TRIM

1956-1962 Corvettes did not have a tag that provided paint and trim code information.

EXTERIOR	COVE	INTERIOR	SOFT TOP
Tuxedo Black	Silver	Red:Turquoise:Black:Blue	Black:White:Blue
Roman Red	White	Red:Black	White:Black
Honduras Maroon	White	Black	Black:White
Ermine White	Silver	Red:Turquoise:	Black:White:Blue Black:Blue
Horizon Blue	White	Blue:Red:Black	Blue:Black:White
Sateen Silver	White	Red:Turquoise:Black:Blue	Black:White:Blue
Tasco Turquoise	White	Turquoise:Black	Black:White:Blue
Cascade Green	White	Black	Black:White:Blue

Cove colors shown were optional, body color standard.
These were factory recommended combinations.

PRODUCTION FIGURES

	MONTHLY	CUMULATIVE
October	1168	1168
November	286	1454
December	605	2059
January	1099	3158
February	1202	4360
March	1351	5711
April	1300	7011
May	1156	8167
June	982	9149
July	697	9846
August	415	10261

RADIATORS

Almost all radiators were copper and contained an identification tag soldered to the top tank. This tag contained the part number 3141674, and the date of manufacture represented by a double digit year code and a single alphabetic month code (Jan=A, Feb=B, etc.). Some 270 and 290HP cars used an aluminum radiator with an expansion tank on top. The part number is 3747516. The identification tag is attached to the tank with screws. The three piece fan shroud was used.

RADIATOR CAPS

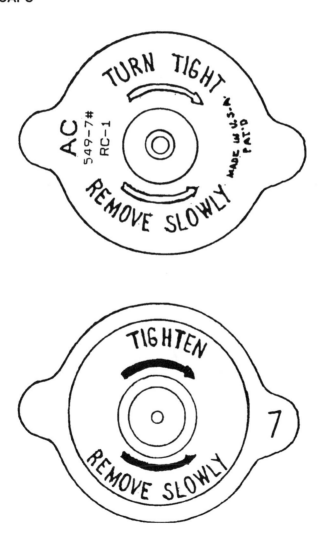

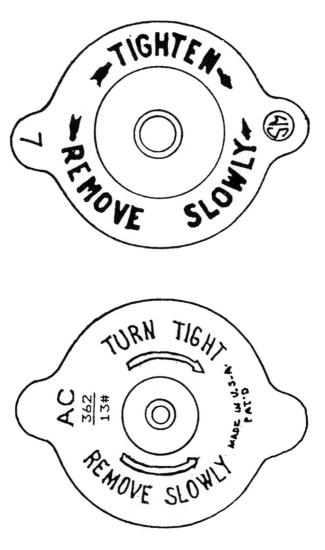

With aluminum radiator

RADIOS

Model number 987730, 7 tube Wonder Bar. The face of the push buttons should be flat. A paper tag on the outside of one of the radio covers contains the model number.

REAR AXLE

Code and build date (month/day) are located on the right, front of the carrier.
CODES
AE .... 3.55, Open, Powerglide
AH .... 3.70, Open, manual
AN .... 3.70, Positraction, manual
AP .... 4.11, Positraction, manual
AQ .... 4.56, Positraction, manual
AS .... 3.70, Positraction, manual, HD brakes & steering
AT .... 4.11, Positraction, manual, HD brakes & steering
AU .... 4.56, Positraction, manual, HD brakes & steering
FJ .... 3.70, Open, manual, metallic brakes
FK .... 3.70, Positraction, manual, metallic brakes
FL .... 4.11, Positraction, manual, metallic brakes
FM .... 4.56, Positraction, manual, metallic brakes

SEATS

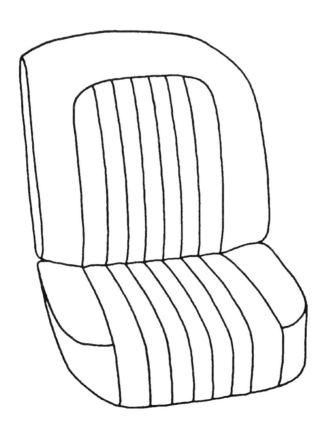

SHIFTERS

The three speed shifter and the automatic transmission shifter have a similar overall appearance even including the mounting bracket. Both utilized a round, thin, chrome plated shift lever. Automatic shift pattern was L D N R P front to back.

The four speed shifter was also somewhat similar even with the addition of the reverse lockout. The shifter tower shaft contained a single spring for gate control tension. This was held in place by a round, slotted nut. An anti rattle spring was attached between the shift lever and a small bracket bolted to the bottom of the shifter support. The reverse lever should be distinctly longer than the third/fourth lever, unlike the 1963 shifters where they are about the same length.

SPARK PLUG WIRES

Distributor boot is 180 degree, black. Spark plug boot is 90 degree, black. Wire is 7MM black with the imprint RADIO GM TVRS, undated.

STARTERS

1107664 was used on all cars. The date and model number were stamped directly into the starter housing. The date is shown as year, month, day, and the month appears as an alphabetic character with A=Jan, B=Feb, etc. The solenoid bakelite should contain Delco Remy.

STEERING WHEELS

Color matching (red, black, turquoise, or blue) wheels were plastic with metal spokes. Outer plastic circumference should be free of grind marks found on reproduction wheels. Finishing marks on the spokes should run the same direction as the rim on the face of the spoke and perpendicular to the rim on the back. The amount of simulated grain, or lack of it, appearing in the plastic of original wheels is a point of controversy. Replacement GM wheels have plenty, but some very original cars have little or none. The idea that the grain has worn away is unlikely since it is typically missing from all areas of the wheel, even areas with little hand contact.

THERMOSTAT HOUSINGS

Engines with a cast iron intake manifold used a curved, cast iron housing, number 3711268. Engines with an aluminum intake manifold used an angled, aluminum housing, number 3837223. The aluminum housing without metal saver notches around the stud bosses is clearly correct for this application. These housings appear in factory photos and on original cars through 1962. Chevrolet provided a slightly different part with the same number as a replacement. Metal saver notches are around the stud bosses and the manufacturers logo appears along with the part number. These may or may not have been used originally.

TIRES

Firestone 'Deluxe Champion', US Royal 'Air Ride', or B.F. Goodrich 'Silvertown' 6.70x15, tubeless. If whitewall equipped, the width should be 2" to 2 9/16" nominal. DOT (department of transportation) information did not appear in the sidewall of original tires.

TRANSMISSIONS

The Borg Warner four speed was available as an option. The aluminum tail housings should not have two half moon cuts at the rear bushing. There should be two shifter mounting positions (5 holes), and M should be cast along with the date. The side cover is dated inside and should contain webbing. The main case is cast iron and has T10-1B and month/day cast into the passenger side. A build code is stamped on the upper rear, driver's side, unmachined part of the case. The code is in the form: location/month/day/year/shift, W=Warner Gear, month=Jan=A, Feb=B, etc., shift=1,2,3. Vehicle serial numbers are not stamped on these transmissions.

Build information is stamped on the upper rear of the case, passenger side for three speeds, and on the rear flange of the governor cover for Powerglide. Build information is in the form: location/month/day/shift (D or N). S=Saginaw (three speed), C=Cleveland (Powerglide). The three speed must have a shifter mounting position on the tail housing.

VALVE COVERS

All valve cover mounting holes were equally spaced at 8 3/4" top and bottom. The 230HP engine used a stamped steel cover with CHEVROLET in raised script.

The remaining engines used a cast aluminum cover with CORVETTE in raised, 1/8" script. A casting flaw should not appear through the 'O' in CORVETTE. All covers had seven fins running the length of the cover. Two notches appeared on the inside lip for intake manifold clearance.

VOLTAGE REGULATORS

1119001, or 1119002 if equipped with 1102173 generator. The cover was stamped with Delco Remy and was attached with slotted cap screws. The part number and date, along with '12VN' for 12 volt negative ground, are stamped in the mounting flange. The date is shown as year/month, with the month represented by an alphabetic character, ie A=Jan, B=Feb, etc.

WATER PUMPS

Casting # 3736493, undated.

WIPER ARMS & BLADES

All cars used a bright stainless, Trico brand, with rubber insert. The Trico name may not always appear on the parts. The rubber inserts have a series of raised dots running the length of both sides. The blade construction consists of a main bridge piece that is somewhat shorter than the overall length of the blade. Attached to this are three links that hold the rubber insert. There are also two hard rubber blocks at each end of the blade.

WIPER MOTORS

5044266 stamped in the front cover. There should not be a ground wire riveted to the rear cover arm. There should be a 1" long metal support stand attached to the lower armature housing bolt. Windshield washers were optional equipment.

1960

```
       JANUARY                    FEBRUARY
S   M   T   W   T   F   S    S   M   T   W   T   F   S
                        1   2                1   2   3   4   5   6
3   4   5   6   7   8   9    7   8   9  10  11  12  13
10  11  12  13  14  15  16  14  15  16  17  18  19  20
17  18  19  20  21  22  23  21  22  23  24  25  26  27
24  25  26  27  28  29  30  28  29
31

        MARCH                       APRIL
S   M   T   W   T   F   S    S   M   T   W   T   F   S
        1   2   3   4   5                                1   2
6   7   8   9  10  11  12    3   4   5   6   7   8   9
13  14  15  16  17  18  19  10  11  12  13  14  15  16
20  21  22  23  24  25  26  17  18  19  20  21  22  23
27  28  29  30  31          24  25  26  27  28  29  30

         MAY                        JUNE
S   M   T   W   T   F   S    S   M   T   W   T   F   S
1   2   3   4   5   6   7                    1   2   3   4
8   9  10  11  12  13  14    5   6   7   8   9  10  11
15  16  17  18  19  20  21  12  13  14  15  16  17  18
22  23  24  25  26  27  28  19  20  21  22  23  24  25
29  30  31                  26  27  28  29  30

         JULY                      AUGUST
S   M   T   W   T   F   S    S   M   T   W   T   F   S
                        1   2       1   2   3   4   5   6
3   4   5   6   7   8   9    7   8   9  10  11  12  13
10  11  12  13  14  15  16  14  15  16  17  18  19  20
17  18  19  20  21  22  23  21  22  23  24  25  26  27
24  25  26  27  28  29  30  28  29  30  31
31

       SEPTEMBER                   OCTOBER
S   M   T   W   T   F   S    S   M   T   W   T   F   S
                    1   2   3                                    1
4   5   6   7   8   9  10    2   3   4   5   6   7   8
11  12  13  14  15  16  17   9  10  11  12  13  14  15
18  19  20  21  22  23  24  16  17  18  19  20  21  22
25  26  27  28  29  30      23  24  25  26  27  28  29
                            30  31

       NOVEMBER                   DECEMBER
S   M   T   W   T   F   S    S   M   T   W   T   F   S
        1   2   3   4   5                        1   2   3
6   7   8   9  10  11  12    4   5   6   7   8   9  10
13  14  15  16  17  18  19  11  12  13  14  15  16  17
20  21  22  23  24  25  26  18  19  20  21  22  23  24
27  28  29  30              25  26  27  28  29  30  31
```

1961

```
       JANUARY                    FEBRUARY
S   M   T   W   T   F   S    S   M   T   W   T   F   S
1   2   3   4   5   6   7                    1   2   3   4
8   9  10  11  12  13  14    5   6   7   8   9  10  11
15  16  17  18  19  20  21  12  13  14  15  16  17  18
22  23  24  25  26  27  28  19  20  21  22  23  24  25
29  30  31                  26  27  28

        MARCH                       APRIL
S   M   T   W   T   F   S    S   M   T   W   T   F   S
            1   2   3   4                                    1
5   6   7   8   9  10  11    2   3   4   5   6   7   8
12  13  14  15  16  17  18   9  10  11  12  13  14  15
19  20  21  22  23  24  25  16  17  18  19  20  21  22
26  27  28  29  30  31      23  24  25  26  27  28  29
                            30

         MAY                        JUNE
S   M   T   W   T   F   S    S   M   T   W   T   F   S
        1   2   3   4   5   6                            1   2   3
7   8   9  10  11  12  13    4   5   6   7   8   9  10
14  15  16  17  18  19  20  11  12  13  14  15  16  17
21  22  23  24  25  26  27  18  19  20  21  22  23  24
28  29  30  31              25  26  27  28  29  30

         JULY                      AUGUST
S   M   T   W   T   F   S    S   M   T   W   T   F   S
                            1            1   2   3   4   5
2   3   4   5   6   7   8    6   7   8   9  10  11  12
9  10  11  12  13  14  15  13  14  15  16  17  18  19
16  17  18  19  20  21  22  20  21  22  23  24  25  26
23  24  25  26  27  28  29  27  28  29  30  31
30  31

       SEPTEMBER                   OCTOBER
S   M   T   W   T   F   S    S   M   T   W   T   F   S
                        1   2   1   2   3   4   5   6   7
3   4   5   6   7   8   9    8   9  10  11  12  13  14
10  11  12  13  14  15  16  15  16  17  18  19  20  21
17  18  19  20  21  22  23  22  23  24  25  26  27  28
24  25  26  27  28  29  30  29  30  31

       NOVEMBER                   DECEMBER
S   M   T   W   T   F   S    S   M   T   W   T   F   S
                1   2   3   4                                1   2
5   6   7   8   9  10  11    3   4   5   6   7   8   9
12  13  14  15  16  17  18  10  11  12  13  14  15  16
19  20  21  22  23  24  25  17  18  19  20  21  22  23
26  27  28  29  30          24  25  26  27  28  29  30
                            31
```

1961 Vehicle Identification Number (VIN) 10867S100001 to 10867S110939, total vehicles 10,939. Plates are located under the hood, spot welded to the steering column.
First digit=model year
S=St. Louis
Six digit sequential serial number starting at 100001

AIR CLEANERS

Both the two four barrel and the single four barrel used a similar louvered air cleaner approximately 14 inches in diameter. There were three rows of louvers in the lightly polished aluminum top. All of the cars had two piece units allowing service of the foam air filter element. The tops were held in place with wing nuts at the carburetor centers. The wing nuts sat in depressions in the top, and the two four barrel unit had an additional depression in the center. Beware of similar, non aluminum, two piece Pontiac air cleaners.

The fuel injection air cleaner is a two piece unit held together with a wing nut. This provided access to the serviceable filter element. The units are all metal and have a large X stamped in the lid that extends out to the very edge of the air cleaner. The units did not bolt directly to the engine, but to the driver's side inner fender.

AIR CONDITIONING

Not Available

ANTENNAS

Assembly 3780070. Three section mast without rings on any section, ball at the top of the top section. Base consists of a round plastic nut with notches for a spanner wrench, a plastic bezel, and a gasket.

BALANCERS

Narrow, 6" diameter balancer with the pulley bolted directly to it.

BATTERIES

1980458 tar top. Year/month and factory of origin is stamped in the top. Cover above the tar on top should be 3 1/2 to 4 inches wide. Caps are all yellow with Delco in black on the outside, visible casting line/flashing on the bottom, inside. Case side should say "Delco DC12 Original Equipment Dry Charge" and should be unpainted (warranty replacements were painted). The characters 459 should be stamped in the side and CAT 458 on top of the battery.

BATTERY CABLES

Negative cable is an uninsulated woven strap with an N on the bolt style terminal connection. Positive cable is black vinyl covered with a P on the bolt style terminal connection.

BELLHOUSINGS

3764591, aluminum without date code.

BRAKE MASTER CYLINDERS

Undated. Part # 5454480 located in the metal plug in the end of the cylinder. Casting # 5456022 located on the driver's side just above the 1" bore size identification. The characters "022" were larger than the others. Filler cap was metal, not plastic, and contains USE GMC BRAKE FLUID around the top.

CARBURETORS

Model numbers are embossed on an aluminum tag attached under one of the carburetor top screws. The date is also shown on this tag in the form month/year, using A=Jan, B=Feb, etc., an additional digit indicated factory production information.

All 230 HP used WCFB 3059S. They have the choke mounted at the bottom, and the choke line is attached to the center of the choke cover. A secondary counterweight is used. The main body is cast 0-1343 or 0-1517, top 6-1396.

All 245 HP used WCFB 2626S front, WCFB 2627S rear early and WCFB 3181S front, WCFB 2627S rear for the remainder. All 270 HP used WCFB 2613S front and

WCFB 2614S rear early and WCFB 3182S front, WCFB 2614S rear for the remainder. All contain secondary counterweights, but no idle air screws. A choke was mounted on the top of the rear carburetor only, and the line was attached at the side, not to the cover. Main bodies are cast 0-049, tops 6-1299 or 0-1049.

CARPET

Salt and pepper appearing loop carpet, foam backed, and sectionally sewn to fit the individual sections and contours of the interior.

COILS

1115091 for carbureted cars, with 091 in raised characters on the case. 1115107 with fuel injection, 107 in raised characters on the case.

CONVERTIBLE TOP

All tops had a rear window attached with heat sealing from the inside. The rear window contained the Vinylite name and trademark, the AS 6 identification, cleaning instructions, and a date code hot stamped into the outside surface. The 1961-1962 top frame differed from earlier frames because of the rear bow, and the filler weatherstrips are consequently longer. The rear bow should match the body contour of the top compartment lid, and is made of aluminum. The straps attach to the rear bow with a clamp, not the earlier tack strip, and a plastic bead is also utilized. The header should be the 59-62 style with a single angle bracket attachment to the side frame. Front hold down latches should have a short handle (1 7/8" across top), not the later 2 1/2" replacement handle. The base of the latch is rounded compared to the tapered angle of the hardtop latch.

CRANKCASE VENTS

Except for the optional PCV, the system consists of the oil filler cap and the road draft tube. The road draft tube between the top and the spark plug grommet holder was primarily straight except where it actually attaches to the motor, the remaining lower area had several slight bends and appears almost curved. See the separate oil cap listing. The optional PCV for California used an adaptor at the rear of the block that looks like the top of the road draft tube. This is connected to the PCV valve, then to the plenum via rubber hose on fuel injected models. The hose is connected directly to the adaptor and runs to the rear of the carburetor where it is attached to the PCV valve on non FI cars.

DISTRIBUTORS

The model used with the 230HP base engine contains a metal tag that wraps around the neck of the distributor below the cap. This contains the date and model number 1110946. All 245 and 270HP cars used model number 1110891 that, along with the date, was stamped on a black and aluminum tag riveted to the distributor housing. The fuel injection cars used models with a cross shaft access cover. The date and model number were stamped in the cover. The 275HP used 1110915. The 315HP used 1110914. Fuel injection distributors all had a facility for driving the tach cable. All models were dated year/month/day.

EMBLEMS

Trunk (Gold Letters)

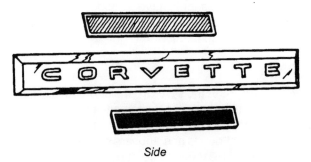

Side

Nose (CORVETTE letters appear under flags)

Side

ENGINE BLOCKS

Casting # 3756519. Very late production casting # 3789935. All engines were 283 cubic inches. Casting number is located on the driver's side rear flange where the bellhousing attaches. The date is located directly opposite on the passenger side rear flange. They are dated month/day/year, with a single digit year representation. Double digits indicate Tonawanda manufacture and are not correct for Corvette. The month is indicated by a consecutive alphabetic character, ie January = A, February = B, etc. A code indicating the assembly location, date, and engine characteristics was stamped on a pad located on the passenger side of the motor. It is just below the cylinder head, next to the water pump. The first character should be F for the Flint assembly plant. The numeric characters represent the date, and the last two characters indicate the engine type as shown in the chart that follows. In 1961 the pad did contain a vehicle serial number, separate from, and to the left of the assembly code.

    CQ = STD 230HP, 4 barrel, manual transmission
    DG = STD 230HP, 4 barrel, automatic transmission
    CT = 469 245HP, 2 four barrels, manual transmission
    DJ = 469 245HP, 2 four barrels, automatic transmission
    CR = 353 275HP, fuel injection, manual transmission
    CU = 468 270HP, 2 four barrels, special cam, manual
    CS = 354 315HP, fuel injection, special cam, manual

EXHAUST MANIFOLDS

Dated month and day only for correct Flint manifolds. Casting # 3749965 left, 3750556 right.

FUEL FILTERS

All cars with two four barrels used a glass sediment bowl style of filter. A 'stone' filter element was used and the bowl was held in place by a wire bail. The primarily flat cast alloy top contained a small AC logo. Cars with one four barrel used a filter mounted in the carburetor just behind the inlet fitting.

Fuel injected cars used a screw together, canister style filter with a removable element. Both fuel lines connected to the sides of the top portion of the assembly, and the unit was mounted behind the fuel meter.

FUEL INJECTION

Four different units were used in 1961. The 7017200 and 7017250 units had a ribbed plenum. The 7017310 and 7017320 had a flat plenum. Both contained a black/aluminum tag showing the part number and a serial number on the driver's side. The serial number has no relationship to any other number. A snowflake was not cast into the plenum top. Cold start enrichment was assisted by an electric thermostatic coil mounted on the air meter. See unit description for additional details. A conventional choke with it's associated butterfly was not used.

7017200   250HP. Cranking signal valve mounted in the plenum was part of the cold enrichment. Balance tube ran from the fuel meter to the air cleaner adaptor. Two lines ran from the main diaphragm, one to the air meter, and one to the cranking signal valve.
Air meter stamped 7017201, cast 7014922
Fuel meter stamped 7017202, cast 7014889

7017250   290HP. Cranking signal valve mounted in the plenum was part of the cold enrichment. Balance tube ran from the fuel meter to the air cleaner adaptor. Two lines ran from the main diaphragm, one to the air meter, and one to the cranking signal valve.
Air meter stamped 7017251, cast 7014922
Fuel meter stamped 7017252, cast 7014889

7017310   250HP. Cranking signal valve mounted in the plenum was part of the cold enrichment. Balance tube ran from the fuel meter to the air cleaner adaptor. Two lines ran from the main diaphragm, one to the air meter, and one to the cranking signal valve.
Air meter stamped 7017201, cast 7014922
Fuel meter stamped 7017202, cast 7014889

7017320   290HP. Cranking signal valve mounted in the plenum was part of the cold enrichment. Balance tube ran from the fuel meter to the air cleaner adaptor. Two lines ran from the main diaphragm, one to the air meter, and one to the cranking signal valve.
Air meter stamped 7017251, cast 7014922
Fuel meter stamped 7017252, cast 7014889

FUEL PUMPS

The AC logo is cast directly in the top of the pump. The four digit Delco number is stamped in the edge of the mounting flange and should be 4656. Pumps are single diaphragm, screwed together assemblies, using short, 3/4" screws.

GAUGES

Fuel/Temp - Battery/Oil combinations on the small gauges contain concave lenses with line indicators on an inner metal face. Pointers are pointed and a flat chrome gauge name plate covers the lower half of the gauge face. Temp gauge should be 220 degree. All tachs are 7000 RPM on a metal face with black background and concave lens. Light green, dark green, yellow, orange, and red (high RPM only) warning areas were used until mid year when the light green area changed to tan. Clocks have a square back and are made by Westclox (stamped on the back). The F for the fast/slow indicator should be on the left. Speedometers are 160 MPH having an inner metal face with lines, and a flat outer face with numbers.

GENERATORS

1102043/30 Amp, all except 315HP
1102173/35 Amp, early 315HP
1102268/35 Amp, late with 315HP

The model number and a date code appear on a red/aluminum tag attached to the generator. The year, month, day, are listed as serial number on the tag, and the month appears as an alphabetic character with A=Jan, B=Feb, etc. All units use a one piece fan-pulley assembly. All 1102043 applications used a 3 5/8" pulley with a 1/2" belt groove and a spacer between the fan and pulley. The 1102173 and 1102268 used a 4" pulley with a 1/2" belt groove. Tach drive assembly is attached to the rear on all but 315HP models.

GLASS

All laminated glass was LOF Safety Plate and, along with the logo, was designated as such by AS1 for windshields and AS2 for side glass. Side and rear Plexiglas was AS4. A two character code (month-year) also appeared indicating the date of manufacture. January=N, February=X, March=L, April=G, May=J, June=I, July=U, August=T, September=A, October=Y, November=C, December=V. The letter U indicates 1960, the letter L 1961. Tinted glass was not used.

GRILLE

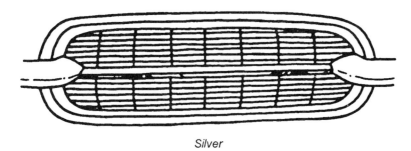

Silver

HARDTOP

Should have a 2" wide, two piece, stainless header trim strip with a small cap covering the center joint. The lower rear stainless trim consists of two pieces joined at the center of the rear window by a curved cap. The curve in this cap is designed to match the top compartment lid. Only two hold down brackets were used on top of the compartment lid, as opposed to the earlier three. Additional hold down brackets were also added on each side just behind the door. All AS-4 Plexiglas windows are dated month/year. Front hold down latches should have a short handle (1 7/8" across top), not the later 2 1/2" replacement handle. The base of the latch is a tapered angle compared to the rounded soft top latch. The white headliner material should be the same grain pattern as the seats, etc.

HEADS

Casting # 3774692 for all carbureted cars, 3782461 for fuel injected cars. The 3782461 heads contain a cast X on the combustion chamber side that is not part of the casting number. Dated month/day/year, with a single digit year representation. Double digits indicate Tonawanda manufacture and are not correct for Corvette. The month is indicated by a consecutive alphabetic character, ie January = A, February = B, etc. Symbol at the end of the head is a partial indication of the application.

HORNS

The last three digits of the part number are stamped into the horn. The high note is 9000442, the low note 9000441. The production date is not stamped. Mounting brackets were welded to the horn, not bolted. These horns differed from 1962 in that they were not dated, and used the thicker, 1960 style mounting bracket.

HORN RELAYS

1116781 for all cars. The characters '781', for the last three digits of the part number, and '12VN', for 12 volt, negative ground, are stamped in the mounting flange. The cover is stamped with Delco Remy.

HUBCAPS

Standard

RPO

IGNITION SHIELDING

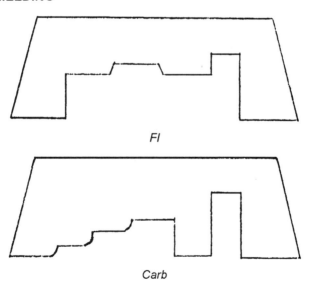

FI

Carb

TOP

All shields were stainless steel and had a front skirt that, except for the center cut-out, was the same height as the rear skirt. All rear attachments are through slots, not holes. The right rear slot should be on an angle parallel to the side. For cars with fuel injection there is a bulge in the front skirt to accommodate the coil.

VERTICALS

All left shields should have the shorter, straight bottom. The left side is stainless, the right is made from chrome plated steel. The right shield lower mounting should be through a hole.

HORIZONTALS

A single piece of shielding is used on each side. Both sides have the smaller exhaust manifold cut-outs, and no reinforcement bar.

INTAKE MANIFOLDS

Dated month/day/year. Casting numbers are located on the top surface. Cast iron manifold date codes are also located on the top surface, and are therefore visible with the manifold installed. Correct manifolds should have a single digit year representation. Aluminum manifold date codes are cast into the bottom surface and are not visible unless the manifold is removed from the engine. Cast iron 230HP manifold was 3746829. The aluminum manifold was 3739653 for both the 245 and 270HP applications.

JACK

Correct scissor jack has an identification number, SJ4653, cast directly into the top of the cast iron base. Jack arms and saddle are stamped steel. Jack screw has fine threads and the handle facility is an oblong slot in the end.

KNOBS

LAMPS

MIRRORS

Outside rearview mirror was stamped 'GUIDE Y-50' on the mirror back at the ball stud. This ball should be centered in the back of the mirror. The glass was not dated, and considered non replaceable since the later style snap ring access was not provided.

Inside mirrors had a base containing a threaded top where the mirror was attached and held in place with a locknut. The mirror was not an exact rectangular shape with the bottom slightly longer than the top. Two rivets surround the ball stud in back, and the GUIDE name is stamped 1/4" below it. The mirror glass was not dated.

OIL CAPS

Solid lifter engines were non vented

OPTIONS

		Quantity	Price
10867	Base convertible	10,939	$3,934.00
Incl.	283, 230HP	5,357	NC
Incl.	3 Speed	2,468	NC
Incl.	Black folding top	3,052	NC
101	Heater	10,671	102.25
102	Wonderbar AM radio	9,316	137.75
242	PCV system		5.40
276	15X5 1/2 Wheels	337	NC
290	Whitewall tires	9,780	31.55
313	Powerglide	1,458	199.10
353	283, 275HP engine	118	484.20
354	283, 315HP engine	1,462	484.20
419	Hardtop	3,395	236.75
419	Hardtop without soft top	2,285	NC
426	Power windows	698	59.20
440	Two tone paint	3,368	16.15
468	283, 270HP engine	2,827	182.95
469	283, 245HP engine	1,175	150.65
470	White soft top	5,602	NC
473	Power folding top	422	161.40
610	Heater delete		NC
675	Positraction rear	6,915	43.05
685	4 Speed transmission	7,013	188.30
686	Metallic brakes	1,402	37.70
687	HD Brakes & steering	233	333.60
1408	Blackwall nylon tires		15.75
1625	24 Gallon fuel tank		161.40

DEALER AVAILABLE OPTIONS
Tool Kit
Spotlight

PAINT/TRIM

1956-1962 Corvettes did not have a tag that provided paint and trim code information.

EXTERIOR	COVE	INTERIOR	SOFT TOP
Tuxedo Black	Silver	Red:Fawn:Black:Blue	Black:White
Roman Red	White	Red:Black	White:Black
Honduras Maroon	White	Black:Fawn	Black:White
Ermine White	Silver	Red:Fawn:Black:Blue	Black:White
Jewel Blue	White	Blue:Black	Black:White
Sateen Silver	White	Red:Black:Blue	Black:White
Fawn Beige	White	Black:Fawn:Red	Black:White

Cove colors shown were optional, body color standard.
These were factory recommended combinations.

PRODUCTION FIGURES

	MONTHLY	CUMULATIVE
September	1052	1052
October	1249	2301
November	1054	3355
December	951	4306
January	897	5203
February	763	5966
March	923	6889
April	915	7804
May	1156	8960
June	1200	10160
July	779	10939

RADIATORS

Some early radiators (about 190 left over from 1960, used until serial number 1000) were copper and contained an identification tag soldered to the top tank. This tag contained the part number 3141674, and the date of manufacture represented by a double digit year code and a single alphabetic month code (Jan=A, Feb=B, etc.). About 1500, 1960 style, aluminum radiators were also used at the beginning of the year. Please see 1960 for details. When this supply was exhausted, an aluminum version with a separate expansion tank was used in all remaining vehicles. The part number was 3150916 for 230, 245, and 275HP, 3151116 for 270 and 315HP. The identification tag was attached to the top of the radiator with screws. Be aware that replacement GM radiators came with a stamped upper water neck, and original radiators used a cast neck. The three piece fan shroud was used.

RADIATOR CAPS

1960 Radiators used 1960 style caps

Used Until March

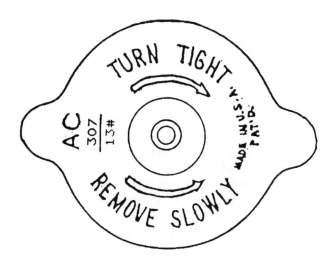

Used After February

RADIOS

Model number 985003, 7 tube Wonder Bar. The face of the push buttons should be curved. Large capacitor can should not be mounted on the back. A paper tag on the outside of one of the radio covers contains the model number.

1961 Corvette

REAR AXLE

Code and build date (month/day) are located on the right, front of the carrier.
CODES
AC .... 3.36, Open, 3 speed
AE .... 3.55, Open, Powerglide
AF .... 3.36, Positraction, manual
AH .... 3.70, Open, 4 speed
AN .... 3.70, Positraction, manual
AP .... 4.11, Positraction, manual
AQ .... 4.56, Positraction, manual
AS .... 3.70, Positraction, manual, HD brakes & steering
AT .... 4.11, Positraction, manual, HD brakes & steering
AU .... 4.56, Positraction, manual, HD brakes & steering
FJ .... 3.70, Open, manual, metallic brakes
FK .... 3.70, Positraction, manual, metallic brakes
FL .... 4.11, Positraction, manual, metallic brakes
FM .... 4.56, Positraction, manual, metallic brakes

SEATS

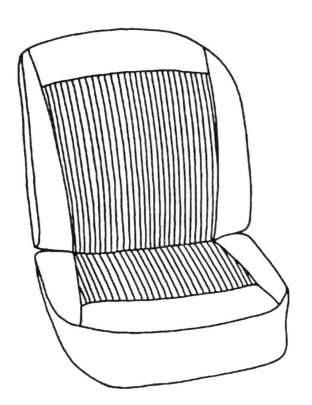

SHIFTERS

The three speed shifter and the automatic transmission shifter have a similar overall appearance even including the mounting bracket. Both utilized a round, thin, chrome plated shift lever. Automatic shift pattern was L D N R P front to back.

The four speed shifter was also somewhat similar even with the addition of the reverse lockout. The shifter tower shaft contained a single spring for gate control tension. This was held in place by a round, slotted nut. An anti rattle spring was attached between the shift lever and a small bracket bolted to the bottom of the shifter support. The reverse lever should be distinctly longer than the third/fourth lever, unlike the 1963 shifters where they are about the same length.

SPARK PLUG WIRES

Distributor boot is 180 degree, black. Spark plug boot is 90 degree, black. All except late cars, wire is 7MM black with the imprint RADIO GM TVRS, undated. Late cars, wire is 7MM with the imprint PACKARD and an indication that the wire is radio/TV suppression, they are dated quarterly. The date appeared as 3-Q-62, for example, as the third quarter 1962.

STARTERS

1107889 was used on all cars. The date and model number were stamped directly into the starter housing. The date is shown as year, month, day, and the month appears as an alphabetic character with A=Jan, B=Feb, etc. The solenoid bakelite should contain Delco Remy.

STEERING WHEELS

Color matching (red, black, fawn, or blue) wheels were plastic with metal spokes. Outer plastic circumference should be free of grind marks found on reproduction wheels. Finishing marks on the spokes should run the same direction as the rim on the face of the spoke and perpendicular to the rim on the back. The amount of simulated grain, or lack of it, appearing in the plastic of original wheels is a point of controversy. Replacement GM wheels have plenty, but some very original cars have little or none. The idea that the grain has worn away is unlikely since it is typically missing from all areas of the wheel, even areas with little hand contact.

THERMOSTAT HOUSINGS

Engines with a cast iron intake manifold used a curved, cast iron housing, number 3711268. Engines with an aluminum intake manifold used an angled, aluminum housing, number 3837223. The aluminum housing without metal saver notches around the stud bosses is clearly correct for this application. These housings appear in factory photos and on original cars through 1962. Chevrolet provided a slightly different part with the same number as a replacement. Metal saver notches are around the stud bosses and the manufacturers logo appears along with the part number. These may or may not have been used originally.

TIRES

Firestone 'Deluxe Champion', US Royal 'Air Ride', or B.F. Goodrich 'Silvertown' 6.70x15, tubeless. If whitewall equipped, the width should be 2" to 2 9/16" nominal. DOT (department of transportation) information did not appear in the sidewall of original tires.

TRANSMISSIONS

The Borg Warner four speed was available as an option. The aluminum tail housings should not have two half moon cuts at the rear bushing. There should be two shifter mounting positions (5 holes), and M should be cast along with the date. The side cover is dated inside until late 1960 after which no dates appear thru 1961. The cover should contain webbing. Until November 1960 the main case is cast iron and has T10-1B and month/day cast into the passenger side, after this the case was T10-1C and was cast aluminum. A build code is stamped on the upper rear, driver's side, unmachined part of the case. The code is in the form: location/month/day/year/shift, W=Warner Gear, month=Jan=A, Feb=B, etc., shift=1,2,3. Vehicle serial numbers are not stamped on these transmissions.

Build information is stamped on the upper rear of the case, passenger side for three speeds, and on the rear flange of the governor cover for Powerglide. Build information is in the form: location/month/day/shift (D or N). S=Saginaw (three speed), C=Cleveland (Powerglide). The three speed must have a shifter mounting position on the tail housing.

VALVE COVERS

All valve cover mounting holes were equally spaced at 8 3/4" top and bottom. The 230HP engine used a stamped steel cover with CHEVROLET in raised script.

The remaining engines used a cast aluminum cover with CORVETTE in raised, 1/8" script. A casting flaw should not appear through the 'O' in CORVETTE. All covers had seven fins running the length of the cover. Two notches appeared on the inside lip for intake manifold clearance.

VOLTAGE REGULATORS

1119001 all except 315HP, 315HP used 1119002. The cover was stamped with Delco Remy and was attached with slotted cap screws. The part number and date, along with '12VN' for 12 volt negative ground, are stamped in the mounting flange. The date is shown as year/month, with the month represented by an alphabetic character, ie A=Jan, B=Feb, etc.

WATER PUMPS

Casting #3782609, undated with a small hole to accept the 5/8"NP bypass fitting without using an adaptor.

WIPER ARMS & BLADES

All cars used a bright stainless, Trico brand, with rubber insert. The Trico name may not always appear on the parts. The rubber inserts have a series of raised dots running the length of both sides. The blade construction consists of a main bridge piece that is somewhat shorter than the overall length of the blade. Attached to this are three links that hold the rubber insert. There are also two hard rubber blocks at each end of the blade.

WIPER MOTORS

5044266 stamped in the front cover. There should not be a ground wire riveted to the rear cover arm. There should be a 1" long metal support stand attached to the lower armature housing bolt. Windshield washers were standard equipment.

1961

```
         JANUARY                    FEBRUARY
S  M  T  W  T  F  S        S  M  T  W  T  F  S
1  2  3  4  5  6  7                    1  2  3  4
8  9 10 11 12 13 14        5  6  7  8  9 10 11
15 16 17 18 19 20 21      12 13 14 15 16 17 18
22 23 24 25 26 27 28      19 20 21 22 23 24 25
29 30 31                  26 27 28

          MARCH                       APRIL
S  M  T  W  T  F  S        S  M  T  W  T  F  S
         1  2  3  4                             1
5  6  7  8  9 10 11        2  3  4  5  6  7  8
12 13 14 15 16 17 18       9 10 11 12 13 14 15
19 20 21 22 23 24 25      16 17 18 19 20 21 22
26 27 28 29 30 31         23 24 25 26 27 28 29
                          30

           MAY                        JUNE
S  M  T  W  T  F  S        S  M  T  W  T  F  S
   1  2  3  4  5  6                       1  2  3
7  8  9 10 11 12 13        4  5  6  7  8  9 10
14 15 16 17 18 19 20      11 12 13 14 15 16 17
21 22 23 24 25 26 27      18 19 20 21 22 23 24
28 29 30 31               25 26 27 28 29 30

          JULY                       AUGUST
S  M  T  W  T  F  S        S  M  T  W  T  F  S
                     1              1  2  3  4  5
2  3  4  5  6  7  8        6  7  8  9 10 11 12
9 10 11 12 13 14 15       13 14 15 16 17 18 19
16 17 18 19 20 21 22      20 21 22 23 24 25 26
23 24 25 26 27 28 29      27 28 29 30 31
30 31

        SEPTEMBER                   OCTOBER
S  M  T  W  T  F  S        S  M  T  W  T  F  S
                  1  2     1  2  3  4  5  6  7
3  4  5  6  7  8  9        8  9 10 11 12 13 14
10 11 12 13 14 15 16      15 16 17 18 19 20 21
17 18 19 20 21 22 23      22 23 24 25 26 27 28
24 25 26 27 28 29 30      29 30 31

         NOVEMBER                   DECEMBER
S  M  T  W  T  F  S        S  M  T  W  T  F  S
         1  2  3  4                             1  2
5  6  7  8  9 10 11        3  4  5  6  7  8  9
12 13 14 15 16 17 18      10 11 12 13 14 15 16
19 20 21 22 23 24 25      17 18 19 20 21 22 23
26 27 28 29 30            24 25 26 27 28 29 30
                          31
```

1962

```
         JANUARY                    FEBRUARY
S  M  T  W  T  F  S        S  M  T  W  T  F  S
   1  2  3  4  5  6                       1  2  3
7  8  9 10 11 12 13        4  5  6  7  8  9 10
14 15 16 17 18 19 20      11 12 13 14 15 16 17
21 22 23 24 25 26 27      18 19 20 21 22 23 24
28 29 30 31               25 26 27 28

          MARCH                       APRIL
S  M  T  W  T  F  S        S  M  T  W  T  F  S
               1  2  3     1  2  3  4  5  6  7
4  5  6  7  8  9 10        8  9 10 11 12 13 14
11 12 13 14 15 16 17      15 16 17 18 19 20 21
18 19 20 21 22 23 24      22 23 24 25 26 27 28
25 26 27 28 29 30 31      29 30

           MAY                        JUNE
S  M  T  W  T  F  S        S  M  T  W  T  F  S
      1  2  3  4  5                             1  2
6  7  8  9 10 11 12        3  4  5  6  7  8  9
13 14 15 16 17 18 19      10 11 12 13 14 15 16
20 21 22 23 24 25 26      17 18 19 20 21 22 23
27 28 29 30 31            24 25 26 27 28 29 30

          JULY                       AUGUST
S  M  T  W  T  F  S        S  M  T  W  T  F  S
1  2  3  4  5  6  7                    1  2  3  4
8  9 10 11 12 13 14        5  6  7  8  9 10 11
15 16 17 18 19 20 21      12 13 14 15 16 17 18
22 23 24 25 26 27 28      19 20 21 22 23 24 25
29 30 31                  26 27 28 29 30 31

        SEPTEMBER                   OCTOBER
S  M  T  W  T  F  S        S  M  T  W  T  F  S
                     1     1  2  3  4  5  6
2  3  4  5  6  7  8        7  8  9 10 11 12 13
9 10 11 12 13 14 15       14 15 16 17 18 19 20
16 17 18 19 20 21 22      21 22 23 24 25 26 27
23 24 25 26 27 28 29      28 29 30 31
30

         NOVEMBER                   DECEMBER
S  M  T  W  T  F  S        S  M  T  W  T  F  S
            1  2  3                             1
4  5  6  7  8  9 10        2  3  4  5  6  7  8
11 12 13 14 15 16 17       9 10 11 12 13 14 15
18 19 20 21 22 23 24      16 17 18 19 20 21 22
25 26 27 28 29 30         23 24 25 26 27 28 29
                          30 31
```

1962 Vehicle Identification Number (VIN) 20867S100001 to 20867S114531, total vehicles 14,531. Plates are located under the hood, spot welded to the steering column.
First digit=model year
S=St. Louis
Six digit sequential serial number starting at 100001

AIR CLEANERS

The single four barrel used a louvered air cleaner approximately 14 inches in diameter. There were three rows of louvers in the lightly polished aluminum top. All of the cars had two piece units allowing service of the foam air filter element. The top was held in place with a wing nut at the carburetor center. The wing nut sat in a depression in the top. Beware of similar, non aluminum, two piece Pontiac air cleaners.

The fuel injection air cleaner is a two piece unit held together with a wing nut. This provided access to the serviceable filter element. The units are all metal and early cars have a large X stamped in the lid that extends out to the very edge of the air cleaner. Most cars had a small X that did not reach the edge of the lid. The units did not bolt directly to the engine, but to the driver's side inner fender.

AIR CONDITIONING

Not Available

ANTENNAS

Assembly 3780070. Three section mast without rings on any section, ball at the top of the top section. Base consists of a round plastic nut with notches for a spanner wrench, a plastic bezel, and a gasket.

BALANCERS

Narrow, 6" diameter balancer with the pulley bolted directly to it was used on all engines with hydraulic lifters. The solid lifter motors used a 1 11/16" wide, 8" diameter balancer with a bolt on pulley. This balancer was dated month/year on the rear face and also contained cooling fins projecting from the inner rear flange.

BATTERIES

1980558 tar top. Year/month and factory of origin is stamped in the top. Cover above the tar on top should be 2 inches wide. Caps are all yellow with Delco in black on the outside, visible casting line/flashing on the bottom, inside. Case side should say "Delco Original Equipment Power Rated" and should be unpainted (warranty replacements were painted). The characters 559 should be stamped in the side and CAT 558 on top of the battery.

BATTERY CABLES

Both cables are spring style and vinyl covered. The part number is stamped near the terminal end. The negative terminal contains an N and the positive a P, but the P will be covered by a short black sleeve. The positive cable is black the negative brown.

BELLHOUSINGS

3779553, aluminum without date code.

BRAKE MASTER CYLINDERS

Undated. Part # 5454480 located in the metal plug in the end of the cylinder. Casting # 5456022 located on the driver's side just above the 1" bore size identification. The characters "022" were larger than the others. Filler cap was plastic, and contains USE DELCO BRAKE FLUID around the top.

CARBURETORS

Model numbers are embossed on an aluminum tag attached under one of the carburetor top screws. The date is also shown on this tag in the form month/year, using A=Jan, B=Feb, etc., an additional digit indicated factory production information. AFB carburetors have the model number and date stamped on the front of the base, passenger side. The date is in the same form as the aluminum tag, but starting in January it is preceded by A, B, or C indicating the production shift.

All 250 HP with manual transmission used WCFB 3191S, with Powerglide WCFB 3190S. They have the choke mounted at the bottom, and the choke line is attached to the center of the choke cover. A secondary counterweight is used. The main body is cast 0-1451 or 0-1465, top 6-1586.

The 300 HP with Powerglide used AFB 3310S. The 300 HP and 340 HP manual transmission cars used AFB 3269S. The main body is cast 0-1481, top 6-1518 for both.

CARPET

Salt and pepper appearing loop carpet, foam backed, and sectionally sewn to fit the individual sections and contours of the interior.

COILS

1115091 for carbureted cars, with 091 in raised characters on the case. 1115107 with fuel injection, 107 in raised characters on the case.

CONVERTIBLE TOP

All tops had a rear window attached with heat sealing from the inside. The rear window contained the Vinylite name and trademark, the AS 6 identification, cleaning instructions, and a date code hot stamped into the outside surface. The 1961-1962 top frame differed from earlier frames because of the rear bow, and the filler weatherstrips are consequently longer. The rear bow should match the body contour of the top compartment lid, and is made of aluminum. The straps attach to the rear bow with a clamp, not the earlier tack strip, and a plastic bead is also utilized. The header should be the 59-62 style with a single angle bracket attachment to the side frame. Front hold down latches should have a short handle (1 7/8" across top), not the later 2 1/2" replacement handle. The base of the latch is rounded compared to the tapered angle of the hardtop latch.

CRANKCASE VENTS

Except for the optional PCV, the system consists of the oil filler cap and the road draft tube. The road draft tube between the top and the spark plug grommet holder was primarily straight except where it actually attaches to the motor, the remaining lower area had several slight bends and appears almost curved. See the separate oil cap listing. The optional PCV for California used an adaptor at the rear of the block that looks like the top of the road draft tube. A rubber hose is connected directly to the adaptor and runs to the rear of the carburetor or plenum where it is attached to the PCV valve.

DISTRIBUTORS

The models used with the carbureted engines contain a metal tag that wraps around the neck of the distributor below the cap. This contains the date and model number. All 250 and 300HP cars used model number 1110984. The 340HP used 1110985. The fuel injection cars used models 1110990 early, and 1111011 after mid year. Both fuel injection models had a cross shaft access cover. The date and model number were stamped in the cover. All models were dated year/month/day. All models also contained a facility for driving the tachometer cable.

EMBLEMS

Rear

Nose (CORVETTE letters appear below flags)

Side

Side (Thin style for 1962)

ENGINE BLOCKS

Casting # 3782870 in characters 1/2" high. Smaller characters were used on passenger car blocks. All engines were 327 cubic inches. Casting number is located on the driver's side rear flange where the bellhousing attaches. The date is located directly opposite on the passenger side rear flange. They are dated month/day/year, with a single digit year representation. Double digits indicate Tonawanda manufacture and are not correct for Corvette. The month is indicated by a consecutive alphabetic character, ie January = A, February = B, etc. A code indicating the assembly location, date, and engine characteristics was stamped on a pad located on the passenger side of the motor. It is just below the cylinder head, next to the water pump. The first character should be F for the Flint assembly plant. The numeric characters represent the date, and the last two characters indicate the engine type as shown in the chart that follows. The pad also contained a vehicle serial number, separate from, and to the left of the assembly code.

    RC = STD 250HP, 4 barrel, manual transmission
    SC = STD 250HP, 4 barrel, automatic transmission
    RD = 583 300HP, 4 barrel, high performance, manual
    SD = 583 300HP, 4 barrel, high performance, automatic
    RE = 396 340HP, 4 barrel, special high performance, manual transmission
    RF = 582 360HP, fuel injection, manual transmission

EXHAUST MANIFOLDS

Casting # 250HP only: 3749965 left, 3750556 right, both dated month and day only for correct Flint manifolds. All other engines 3797901 left, 3797902 right, both were originally undated.

FUEL FILTERS

All cars with AFB carburetors used a glass sediment bowl style of filter. A 'stone' filter element was used and the bowl was held in place by a wire bail. The primarily flat cast alloy top contained a small AC logo. Cars with a WCFB four barrel used a filter mounted in the carburetor just behind the inlet fitting.

Early fuel injected cars with the 7017355 used a screw together, canister style filter with a removable element. Both fuel lines connected to the sides of the top portion of the assembly, and the unit was mounted behind the fuel meter. The 7017360 units all used the disposable GF 90 inline filter mounted directly to the back of the fuel meter. The color of the filter should be silver.

FUEL INJECTION

Two units were used in 1962, the only difference being the style of fuel filter. Very few of the 7017355 units were used at the start of production and these utilized the earlier canister type fuel filter. All 7017360 units used the inline, throw away type filter. Both had a flat plenum that contained a black/aluminum tag showing the part number and a serial number on the driver's side. The serial number has no relationship to any other number. All chokes were electric with a conventional butterfly in the air meter. The cranking signal valve is located on the enrichment diaphragm and connects to one side of the main diaphragm T. The other side of the T connects to the air meter. The enrichment diaphragm line connects to the front of the plenum. The forward balance tube runs from the fuel meter to the air cleaner adaptor.

7017355 and 7017360 360HP
    Air meter is unstamped, cast 7014922
    Fuel meter 7017252 is unstamped, cast 7014889

FUEL PUMPS

The AC logo is cast directly in the top of the pump. The four digit Delco number is stamped in the edge of the mounting flange and should be 4656. Pumps are single diaphragm, screwed together assemblies, using short, 3/4" screws.

GAUGES

Fuel/Temp - Battery/Oil combinations on the small gauges contain concave lenses with line indicators on a metal face. Pointers are pointed and a flat chrome gauge name plate covers the lower half of the gauge face. The temperature gauge changed from 220 to 240 degree early in 1962. The 240 degree was used as a service replacement for earlier years. All tachs are 7000 rpm on a metal face containing tan, dark green, yellow, orange, and red (high rpm only) warning areas, and an outer concave lens. Clocks have a square back and are made by Westclox (stamped on the back). The F for the slow/fast indicator should be on the left. Speedometers are 160 MPH having an inner metal face with lines, and a flat outer face with numbers.

GENERATORS

    1102174/35 Amp, all without solid lifters
    1102268/35 Amp, all with solid lifters

The model number and a date code appear on a red/aluminum tag attached to the generator. The year, month, day, are listed as serial number on the tag, and the month appears as an alphabetic character with A=Jan, B=Feb, etc. All units use a one piece fan-pulley assembly. All 1102174 applications used a 3 5/8" pulley with a 1/2" belt groove and a spacer between the fan and pulley. The 1102268 used a 4" pulley with a 1/2" belt groove. Neither unit used a tach drive assembly.

GLASS

All laminated glass was LOF Safety Plate and, along with the logo, was designated as such by AS1 for windshields and AS2 for side glass. Side and rear Plexiglas was AS4. A two character code (month-year) also appeared indicating the date of manufacture. January=N, February=X, March=L, April=G, May=J, June=I, July=U, August=T, September=A, October=Y, November=C, December=V. The letter L indicates 1961, the letter I 1962. Tinted glass was not used.

GRILLE

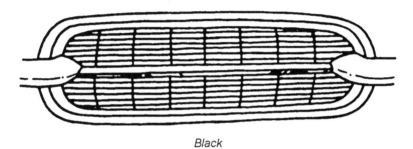

Black

HARDTOP

Should have a 2" wide, two piece, stainless header trim strip with a small cap covering the center joint. The lower rear stainless trim consists of two pieces joined at the center of the rear window by a curved cap. The curve in this cap is designed to match the top compartment lid. Only two hold down brackets were used on top of the compartment lid, as opposed to the earlier three. Additional hold down brackets were also added on each side just behind the door. All AS-4 Plexiglas windows are dated month/year. Front hold down latches should have a short handle (1 7/8" across top), not the later 2 1/2" replacement handle. The base of the latch is a tapered angle compared to the rounded soft top latch. The white headliner material should be the same grain pattern as the seats, etc.

HEADS

Casting # 3795896 for all 250HP cars, 3782461 for all others. The 3782461 heads contain a cast X on the combustion chamber side that is not part of the casting number. Dated month/day/year, with a single digit year representation. Double digits indicate Tonawanda manufacture and are not correct for Corvette. The month is indicated by a consecutive alphabetic character, ie January = A, February = B, etc. Symbol at the end of the head is a partial indication of the application.

HORNS

The last three digits of the part number are stamped into the horn. The high note is 9000442, the low note 9000441. The production date is also stamped year/month/week (1-5). Mounting brackets were welded to the horn, not bolted. These horns differed from 1961 in that they were dated, and used a thinner mounting bracket.

HORN RELAYS

1116781 for all cars. The characters '781', for the last three digits of the part number, and '12VN', for 12 volt, negative ground, are stamped in the mounting flange. The cover is stamped with Delco Remy.

HUBCAPS

Standard

RPO

IGNITION SHIELDING

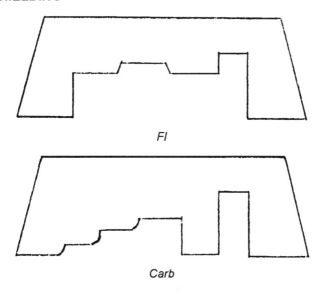

FI

Carb

TOP

All shields were stainless steel and had a front skirt that, except for the center cut-out, was the same height as the rear skirt. All rear attachments are through slots, not holes. The right rear slot should be straight up and down, perpendicular to the top. For cars with fuel injection there is a bulge in the front skirt to accommodate the coil.

VERTICALS

All left shields should have the shorter, straight bottom. The left side is stainless, the right is made from chrome plated steel. The right shield lower mounting should be through a slot.

HORIZONTALS

A single piece of shielding is used on each side. Both sides have the larger exhaust manifold cut-outs, and a reinforcement bar.

INTAKE MANIFOLDS

Dated month/day/year. Casting numbers are located on the top surface. Cast iron manifold date codes are also located on the top surface, and are therefore visible with the manifold installed. Correct manifolds should have a single digit year representation. Aluminum manifold date codes are cast into the bottom surface and are not visible unless the manifold is removed from the engine. Cast iron 250HP manifold was 3783244. Cast iron 300HP manifold was 3799349. The 340HP aluminum manifold was 3795397.

JACK

Correct scissor jack has a base, arms, and saddle made of stamped steel. Manufacturer's logo, 'A', is stamped in the base. Jack screw has coarse threads and the handle facility is a hole in the end cross drilled to accept the handle tang.

KNOBS

LAMPS

MIRRORS

Outside rearview mirror was stamped 'GUIDE Y-50' on the mirror back at the ball stud. This ball should be centered in the back of the mirror. The glass was not dated, and considered non replaceable since the later style snap ring access was not provided.

Inside mirrors had a base containing a threaded top where the mirror was attached and held in place with a locknut. The mirror was not an exact rectangular shape with the bottom slightly longer than the top. Two rivets surround the ball stud in back, and the GUIDE name is stamped 1/4" below it. The mirror glass was not dated.

OIL CAPS

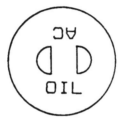

Solid lifter engines were non vented

OPTIONS

		Quantity	Price
20867	Base convertible	14,531	$4,038.00
Incl.	327, 250HP	4,907	NC
Incl.	3 Speed	1,681	NC
Incl.	Black folding top	4,727	NC
102	Wonderbar AM radio	13,076	137.75
242	PCV system		5.40
276	15X5 1/2 Wheels	561	NC
313	Powerglide	1,532	199.10
396	327, 340HP engine	4,412	107.60
419	Hardtop	4,895	236.75
419	Hardtop without soft top	3,179	NC
426	Power windows	995	59.20
470	White soft top	6,625	NC
473	Power folding top	350	139.90
488	24 Gallon fuel tank	65	118.40
582	327, 360HP engine	1,918	484.20
583	327, 300HP engine	3,294	53.80
675	Positraction rear	14,232	43.05

685	4 Speed transmission	11,318	188.30
686	Metallic brakes	2,799	37.70
687	HD Brakes & steering	246	333.60
1832	Whitewall tires		31.55
1833	Blackwall nylon tires		15.70

DEALER AVAILABLE OPTIONS
Tool Kit
Spotlight

PAINT/TRIM

1956-1962 Corvettes did not have a tag that provided paint and trim code information.

EXTERIOR	INTERIOR	SOFT TOP
Tuxedo Black	Red:Fawn:Black	Black:White
Roman Red	Red:Black:Fawn	White:Black
Honduras Maroon	Black:Fawn	Black:White
Ermine White	Red:Fawn:Black	Black:White
Almond Beige	Red:Fawn	Black:White
Sateen Silver	Red:Black	Black:White
Fawn Beige	Fawn:Red	Black:White

These were factory recommended combinations.

PRODUCTION FIGURES

	MONTHLY	CUMULATIVE
August	443	443
September	384	827
October	1238	2065
November	1400	3465
December	1301	4766
January	1468	6234
February	1351	7585
March	1531	9116
April	1403	10519
May	1516	12035
June	1424	13459
July	1061	14520
August	11	14531

RADIATORS

All radiators were aluminum, part number 3150916, with a separately mounted expansion tank. Most radiators contained a metal tag screwed to the top that showed the part number and date of manufacture represented by a double digit year code and a single alphabetic month code (Jan=A, Feb=B, etc.). Later in the year this information was stamped directly into the top of the radiator. Be aware that replacement GM radiators came with a stamped upper water neck, and original radiators used a cast neck. The three piece fan shroud was used.

RADIATOR CAPS

RADIOS

Model number 987383, 7 tube Wonder Bar. The face of the push buttons should be curved. Large capacitor can should be mounted on the back. A paper tag on the outside of one of the radio covers contains the model number.

REAR AXLE

Code and build date (month/day) are located on the right, front of the carrier.
CODES
CA .... 3.36, Open, all
CB .... 3.36, Positraction, all
CC .... 3.55, Positraction, 4 speed
CD .... 3.70, Positraction, 4 speed
CE .... 4.11, Positraction, 4 speed
CF .... 4.56, Positraction, 4 speed
CG .... 3.70, Open, 4 speed
CH .... 3.36, Open, manual, metallic brakes
CK .... 3.36, Positraction, 4 speed, metallic brakes
CL .... 3.55, Positraction, 4 speed, metallic brakes

CM .... 3.70, Positraction, 4 speed, metallic brakes
CN .... 4.11, Positraction, 4 speed, metallic brakes
CP .... 4.56, Positraction, 4 speed, metallic brakes
CQ .... 3.70, Positraction, 4 speed, HD brakes & steering
CR .... 4.11, Positraction, 4 speed, HD brakes & steering
CS .... 4.56, Positraction, 4 speed, HD brakes & steering
CT .... 3.08, Open, 4 speed
CU .... 3.08, Positraction, 4 speed
CV .... 3.08, Open, 4 speed, metallic brakes
CW .... 3.08, Positraction, 4 speed, metallic brakes

SEATS

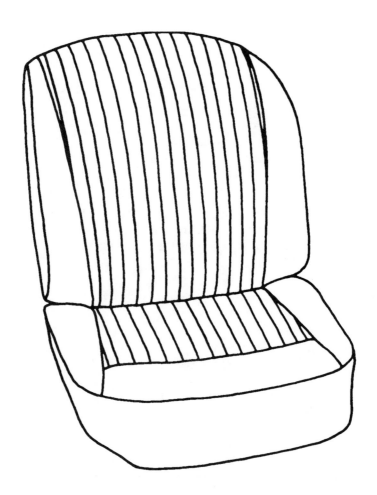

SHIFTERS

The three speed shifter and the automatic transmission shifter have a similar overall appearance, even including the mounting bracket. Both utilized a round, thin, chrome plated shift lever. Automatic shift pattern was P R N D L front to back.

The four speed shifter was also somewhat similar even with the addition of the reverse lockout. The shifter tower shaft contained a single spring for gate control tension. This was held in place by a round, slotted nut. An anti rattle spring was attached between the shift lever and a small bracket bolted to the bottom of the shifter support. The reverse lever should be distinctly longer than the third/fourth lever, unlike the 1963 shifters where they are about the same length.

SPARK PLUG WIRES

Distributor boot is 180 degree, black. Spark plug boot is 90 degree, black. Wire is 7MM with the imprint PACKARD and an indication that they are radio/TV suppression, they are dated quarterly. The date appeared as 3-Q-62, for example, as the third quarter 1962.

STARTERS

1107233 all except Powerglide, 1107242 with Powerglide. The date and model number were stamped directly into the starter housing. The date is shown as year, month, day, and the month appears as an alphabetic character with A=Jan, B=Feb, etc. The solenoid bakelite should contain Delco Remy.

STEERING WHEELS

Color matching (red, black or fawn) wheels were plastic with metal spokes. Outer plastic circumference should be free of grind marks found on reproduction wheels. Finishing marks on the spokes should run the same direction as the rim on the face of the spoke and perpendicular to the rim on the back. The amount of simulated grain, or lack of it, appearing in the plastic of original wheels is a point of controversy. Replacement GM wheels have plenty, but some very original cars have little or none. The idea that the grain has worn away is unlikely since it is typically missing from all areas of the wheel, even areas with little hand contact.

THERMOSTAT HOUSINGS

Engines with a cast iron intake manifold used a curved, cast iron housing, number 3711268. Engines with an aluminum intake manifold used an angled, aluminum housing, number 3837223. The aluminum housing without metal saver notches around the stud bosses is clearly correct for this application. These housings appear in factory photos and on original cars through 1962. Chevrolet provided a slightly different part with the same number as a replacement. Metal saver notches are around the stud bosses and the manufacturers logo appears along with the part number. These may or may not have been used originally.

TIRES

General 'Jet Air', Goodyear 'Custom Super Cushion', Firestone 'Deluxe Champion', US Royal 'Air Ride', or B.F. Goodrich 'Silvertown' 6.70x15, tubeless. If whitewall equipped, the width should be 7/8" to 1 1/16" nominal. DOT (department of transportation) information did not appear in the sidewall of original tires.

TRANSMISSIONS

The Borg Warner four speed was available as an option. The aluminum tail housings should not have two half moon cuts at the rear bushing. There should be two shifter mounting positions (5 holes), and T10 7D should be cast along with the date. A new forward motor mount position was used on the 1962 7D tailhousing requiring an adaptor plate to attach to the crossmember. The side cover is undated until late 1961 after which dates appear on the outside. The cover should contain webbing. The main case is T10-1C and was cast aluminum. The cast date appeared within circles on the passenger side. A build code is stamped on the upper rear, driver's side, unmachined part of the case. The code is in the form: location/month/day/year/shift, W=Warner Gear, month=Jan=A, Feb=B, etc., shift=1,2,3. Vehicle serial numbers are stamped on these transmissions next to the build information.

Build information is stamped on the upper rear of the case, passenger side for three speeds. For Powerglide until November 1961 the stamping was on the front of the case, passenger side, after that they were stamped in the center of the oil pan. Build information is in the form: location/month/day/shift (D or N). S=Saginaw (three speed), B=Toledo (Powerglide). The three speed must have a shifter mounting position on the tail housing.

VALVE COVERS

All valve cover mounting holes were equally spaced at 8 3/4" top and bottom. The 250 and 300HP engine used a stamped steel cover with a rectangular raised area in the center. The raised area contained a label that read CHEVROLET at the top, 327 in the center, and TURBO-FIRE at the bottom. The sides of the raised area should be straight, not triangular.

The remaining engines used a cast aluminum cover with CORVETTE in raised, 1/8" script. A casting flaw should not appear through the 'O' in CORVETTE. All covers had seven fins running the length of the cover. Two notches appeared on the inside lip for intake manifold clearance.

VOLTAGE REGULATORS

1119002 all cars. The cover was stamped with Delco Remy and was attached with slotted cap screws. The part number and date, along with '12VN' for 12 volt negative ground, are stamped in the mounting flange. The date is shown as year/month, with the month represented by an alphabetic character, ie A=Jan, B=Feb, etc.

WATER PUMPS

Casting #3782609, undated with a small hole to accept the 5/8"NP bypass fitting without using an adaptor.

WIPER ARMS & BLADES

All cars used a bright stainless, Trico brand, with rubber insert. The Trico name may not always appear on the parts. The rubber inserts have a series of raised dots running the length of both sides. The blade construction consists of a main bridge piece that is somewhat shorter than the overall length of the blade. Attached to this are three links that hold the rubber insert. There are also two hard rubber blocks at each end of the blade.

WIPER MOTORS

5044479 stamped in the front cover. The 5044266 motor may have been used early on. There should not be a ground wire riveted to the rear cover arm. There should be a 1" long metal support stand attached to the lower armature housing bolt. Windshield washers were standard equipment.

1962

```
        JANUARY                    FEBRUARY
S   M   T   W   T   F   S     S   M   T   W   T   F   S
    1   2   3   4   5   6                     1   2   3
7   8   9  10  11  12  13     4   5   6   7   8   9  10
14  15  16  17  18  19  20    11  12  13  14  15  16  17
21  22  23  24  25  26  27    18  19  20  21  22  23  24
28  29  30  31                25  26  27  28

         MARCH                       APRIL
S   M   T   W   T   F   S     S   M   T   W   T   F   S
                1   2   3     1   2   3   4   5   6   7
4   5   6   7   8   9  10     8   9  10  11  12  13  14
11  12  13  14  15  16  17    15  16  17  18  19  20  21
18  19  20  21  22  23  24    22  23  24  25  26  27  28
25  26  27  28  29  30  31    29  30

          MAY                        JUNE
S   M   T   W   T   F   S     S   M   T   W   T   F   S
        1   2   3   4   5                             1   2
6   7   8   9  10  11  12     3   4   5   6   7   8   9
13  14  15  16  17  18  19    10  11  12  13  14  15  16
20  21  22  23  24  25  26    17  18  19  20  21  22  23
27  28  29  30  31            24  25  26  27  28  29  30

          JULY                      AUGUST
S   M   T   W   T   F   S     S   M   T   W   T   F   S
1   2   3   4   5   6   7                 1   2   3   4
8   9  10  11  12  13  14     5   6   7   8   9  10  11
15  16  17  18  19  20  21    12  13  14  15  16  17  18
22  23  24  25  26  27  28    19  20  21  22  23  24  25
29  30  31                    26  27  28  29  30  31

       SEPTEMBER                   OCTOBER
S   M   T   W   T   F   S     S   M   T   W   T   F   S
                        1         1   2   3   4   5   6
2   3   4   5   6   7   8     7   8   9  10  11  12  13
9  10  11  12  13  14  15     14  15  16  17  18  19  20
16  17  18  19  20  21  22    21  22  23  24  25  26  27
23  24  25  26  27  28  29    28  29  30  31
30

        NOVEMBER                  DECEMBER
S   M   T   W   T   F   S     S   M   T   W   T   F   S
                1   2   3                             1
4   5   6   7   8   9  10     2   3   4   5   6   7   8
11  12  13  14  15  16  17    9  10  11  12  13  14  15
18  19  20  21  22  23  24    16  17  18  19  20  21  22
25  26  27  28  29  30        23  24  25  26  27  28  29
                              30  31
```

1963

```
         JANUARY                   FEBRUARY
S   M   T   W   T   F   S     S   M   T   W   T   F   S
        1   2   3   4   5                             1   2
6   7   8   9  10  11  12     3   4   5   6   7   8   9
13  14  15  16  17  18  19    10  11  12  13  14  15  16
20  21  22  23  24  25  26    17  18  19  20  21  22  23
27  28  29  30  31            24  25  26  27  28

          MARCH                      APRIL
S   M   T   W   T   F   S     S   M   T   W   T   F   S
                        1   2     1   2   3   4   5   6
3   4   5   6   7   8   9     7   8   9  10  11  12  13
10  11  12  13  14  15  16    14  15  16  17  18  19  20
17  18  19  20  21  22  23    21  22  23  24  25  26  27
24  25  26  27  28  29  30    28  29  30
31

           MAY                        JUNE
S   M   T   W   T   F   S     S   M   T   W   T   F   S
            1   2   3   4                                 1
5   6   7   8   9  10  11     2   3   4   5   6   7   8
12  13  14  15  16  17  18    9  10  11  12  13  14  15
19  20  21  22  23  24  25    16  17  18  19  20  21  22
26  27  28  29  30  31        23  24  25  26  27  28  29
                              30

          JULY                       AUGUST
S   M   T   W   T   F   S     S   M   T   W   T   F   S
    1   2   3   4   5   6                     1   2   3
7   8   9  10  11  12  13     4   5   6   7   8   9  10
14  15  16  17  18  19  20    11  12  13  14  15  16  17
21  22  23  24  25  26  27    18  19  20  21  22  23  24
28  29  30  31                25  26  27  28  29  30  31

       SEPTEMBER                    OCTOBER
S   M   T   W   T   F   S     S   M   T   W   T   F   S
1   2   3   4   5   6   7                 1   2   3   4   5
8   9  10  11  12  13  14     6   7   8   9  10  11  12
15  16  17  18  19  20  21    13  14  15  16  17  18  19
22  23  24  25  26  27  28    20  21  22  23  24  25  26
29  30                        27  28  29  30  31

        NOVEMBER                   DECEMBER
S   M   T   W   T   F   S     S   M   T   W   T   F   S
                        1   2     1   2   3   4   5   6   7
3   4   5   6   7   8   9     8   9  10  11  12  13  14
10  11  12 -13  14  15  16    15  16  17  18  19  20  21
17  18  19  20  21  22  23    22  23  24  25  26  27  28
24  25  26  27  28  29  30    29  30  31
```

1963 Corvette

1963 Vehicle Identification Number (VIN) 308x7S100001 to 308x7S121513, total vehicles 21,513, 10,919 convertibles, 10,594 coupes. Plates are located under the glove compartment, spot welded to the horizontal brace.
First digit=model year
Fourth digit=6 for convertible, 3 for coupe
S=St. Louis
Six digit sequential serial number starting at 100001

AIR CLEANERS

The 250 and 300HP engines used a dual snorkel style air cleaner. The lid was chrome plated and removable to allow servicing of the original foam element. The base was painted black and had a small, approximately 3/8", tube extending from the front to facilitate the hose connection to the oil filler tube. Cars with air conditioning had pieces attached to the carburetor cutout in order to shift the snorkels away from the radiator hose. The 340HP air cleaner was basically the same except the base was chrome plated, and had cutouts in the rear for air intake replacing the snorkels in the front.

The fuel injection air cleaner, and S tube connection to the core support were originally metal. The main canister was mounted to the driver's side inner fender. An elbow was attached to the main canister with three latch type clamps allowing access to the filter element. The uppermost of these three clamps was mounted in front of, not on top of, the fender mounting flange.

AIR CONDITIONING

This was the first year for this option, and only 278 cars were produced. Actual production did not begin until late in the year. Extensive changes were required, both small and large, to accommodate the additional equipment necessary. Major additional items included the condenser, mounted in front of the radiator; the evaporator and case, mounted on the passenger side firewall; the dehydrator (drier), mounted on the core support; the compressor, mounted on the engine, passenger side; two additional dash controls; interior ducting consisting of a small outlet above the clock, and two large outlets mounted under the dash at both ends. Adding these parts required different passenger side horn positioning, a new windshield washer reservoir (a bag instead of a plastic jar), battery relocation to the driver's side, moving the headlight motor switch, a longer hood release bracket. and a different trip odometer cable bracket, among other obvious things such as wiring, etc.

The parts themselves evolved into different versions depending on design and manufacturing changes. Some of the 1963 parts were unique as first year items. The compressor mounting was accomplished using the same passenger side exhaust manifold as non air conditioned cars, and the brackets were all stamped steel. The compressor itself had two metal identification tags on the case with model and serial number, and the rear of the case was dated month/day/year. The month was represented by a letter, Jan=A, Feb=B, etc.

The suction throttling valve, or STV was used to control evaporator pressure from 1963 to 1966, and is mounted on the evaporator case. Both large lines should enter at the top, not top and bottom as seen on POA valves used as replacements.

Original water control valves have a hexagonal portion as part of their construction that was not retained by the later replacements.

Dehydrators (driers) were originally manufactured with a single continuous diameter, later replacements look like an old milk bottle with a smaller diameter at the top.

Replacement or reproduction condensers are more of a universal type having mounting flanges that, unlike the originals, accommodate a number of applications.

All three ducts were unique to 1963. The two end ducts used an air deflector ball without a center divider, and the bezel was attached with screws directly opposite each other. The center duct did not have the dash control standoffs used from 1964 thru 1967 to facilitate dash lamp usage.

ALTERNATORS

  1100628/37 Amp, all without air conditioning
  1100633/52 Amp, with air conditioning

All units are stamped with the amperage rating, the model number, and the date. The date is represented year/month/day, with a letter code used for the month, ie A=Jan, B=Feb, etc. Stampings, after December 1962, were on the front case. Prior to this they were on the rear case. The rear case should NOT have a hole for a wiring harness connector since a clip was bolted to the case for this purpose. All blades on the fan should be the same size, and both the fan and pulley should be cad plated. A metal disc was used between the fan and pulley on air conditioned cars. Cars with solid lifters used a wide, 3 5/8" diameter pulley with a 1/2" belt groove. Cars with hydraulic lifters used a narrow 2 7/8" diameter pulley with a 3/8" belt groove.

ANTENNAS

Assembly 3832327. Three section mast without rings on any section, ball at the top of the top section. Base consists of a round chrome nut with notches for a spanner wrench, a chrome bezel, and a gasket. The fact that the top two sections must collapse completely into the bottom section in order for the antenna to be authentic is erroneous.

BALANCERS

Narrow, 6" diameter balancer with the pulley bolted directly to it was used on all engines with hydraulic lifters. The solid lifter motors used a 1 11/16" wide, 8" diameter balancer with a bolt on pulley. This balancer was dated month/year on the rear face and also contained cooling fins projecting from the inner rear flange.

BATTERIES

1980558 tar top. Year/month and factory of origin is stamped in the top. Cover above the tar on top should be 2 inches wide. Caps are all yellow with Delco in black until mid '63 then black with Delco in yellow. Caps have visible casting line/flashing on the bottom, inside. Case side should say "Delco Original Equipment Power Rated" and should be unpainted (warranty replacements were painted). The characters 559 should be stamped in the side and CAT 558 on top of the battery.

BATTERY CABLES

Except for cars equipped with air conditioning both cables are spring style and vinyl covered. The part number is stamped near the terminal end. The negative terminal contains an N and the positive a P, but the P will be covered by a short black sleeve. The positive cable is black the negative brown. Air conditioned cars used an uninsulated woven strap for the negative cable and a black spring style positive cable.

BELLHOUSINGS

3788421, dated month-day in a small circle usually not readable. Note that original 1963 manual transmissions used a different front bearing retainer than later years. This retainer is smaller and will not work correctly with the 3858403 bellhousing.

BRAKE MASTER CYLINDERS

Dated month/day/year under bowl. With the exception of Z06 equipped cars, the 7/8" bore size identification should be cast into the top of the cylinder at the front (passenger cars used 1"). The cap was actually a small and a large cap held in place by a thumb screw. The large cap contained USE DELCO BRAKE FLUID FILL TO 1/4" BELOW RIM stamped around the edge. Z06 equipped cars used a dual line master cylinder with a screw on cap.

CARBURETORS

Model numbers are embossed on an aluminum tag attached under one of the carburetor top screws. The date is also shown on this tag in the form month/year, using A=Jan, B=Feb, etc., an additional digit indicated factory production information. AFB carburetors have the model number and date stamped on the front of the base, passenger side. The date is in the same form as the aluminum tag, but it is preceded by A, B, or C indicating the production shift. The A,B,C was not used the entire year, but at least through February.

All 250 HP with manual transmission used WCFB 3501S, with Powerglide WCFB 3500S. They have the choke mounted at the bottom, and the choke line is attached to the center of the choke cover. A secondary counterweight is used. The main body is cast 0-1465, top 6-1672.

The 300 HP with Powerglide used AFB 3460S. The 300 HP and 340 HP manual transmission cars used AFB 3461S. The main body is cast 0-1552, top 6-1518 for both.

CARPET

Loop carpet, foam backed, and sectionally sewn to fit the individual sections and contours of the interior.

COILS

1115091 early, or 1115087. The last 3 digits of the part number appear in raised characters on the case.

CONVERTIBLE TOP

All tops had a rear window attached with heat sealing from the inside. The rear window contained the Vinylite name and trademark, the AS 6 identification, cleaning instructions, and a date code hot stamped into the outside surface. The top frame differed in one respect from 1963 to 1967. The first bow, directly above the driver, was round from 1963 to mid 1965. The pads attached to this bow with screws. The bow used from mid 1965 to 1967 was flat, and the pads were attached to a tack strip.

CRANKCASE VENTS

This was the first year for the production 'closed' crankcase ventilation system. The rear engine block opening contained a rubber grommet with a hole in the center. A metal tube ran forward from this grommet where a piece of rubber hose connected it to the PCV valve installed in the rear of the carburetor or the right rear side of the fuel injection plenum. The oil filler tube had a small pipe that was connected to the air cleaner base via a rubber hose, or to the fuel injection adaptor with rubber hoses and a metal tube.

DISTRIBUTORS

The model used with the carbureted engines contains a metal tag that wraps around the neck of the distributor below the cap. This contains the date and model number. All 250, 300 and 340HP cars used model number 1111024. The fuel injection cars used model 1111022. Fuel injection models had a cross shaft access cover. The date and model number were stamped in the cover. All models were dated year/month/day. All models also contained a facility for driving the tachometer cable.

EMBLEMS

Rear Deck

Nose

Side

Side

ENGINE BLOCKS

Casting # 3782870 in characters 1/2" high. Smaller characters were used on passenger car blocks. All engines were 327 cubic inches. Casting number is located on the driver's side rear flange where the bellhousing attaches. The date is located directly opposite on the passenger side rear flange. They are dated month/day/year, with a single digit year representation. Double digits indicate Tonawanda manufacture and are not correct for Corvette. The month is indicated by a consecutive alphabetic character, ie January = A, February = B, etc. A code indicating the assembly location, date, and engine characteristics was stamped on a pad located on the passenger side of the motor. It is just below the cylinder head, next to the water pump. The first character should be F for the Flint assembly plant. The numeric characters represent the date, and the last two characters indicate the engine type as shown in the chart that follows. The pad also contained a vehicle serial number, separate from, and to the left of the assembly code.

RC = STD 250HP, 4 barrel, manual transmission
SC = STD 250HP, 4 barrel, automatic transmission
RD = L75 300HP, 4 barrel, high performance, manual
SD = L75 300HP, 4 barrel, high performance, automatic
RE = L76 340HP, 4 barrel, special high performance, manual transmission
RF = L84 360HP, fuel injection, manual transmission

EXHAUST MANIFOLDS

Casting # for all 250HP, and 300HP with automatic transmission: 3749965 left, 3750556 right, both dated month and day only for correct Flint manifolds. All other carbureted engines 3797901 left, 3797902 right, both originally undated. Fuel injected engines use 3797942 left, and 3797902 right, both originally undated.

FUEL FILTERS

Cars with a WCFB four barrel used a filter mounted in the carburetor just behind the inlet fitting. All other cars used an external GF 90 filter. The color of the GF 90 was initially silver, but changed to black at some point around mid production.

FUEL INJECTION

Only one unit was used in 1963 and it used a wide, removable, ribbed plenum cover. The plenum contained a black/aluminum tag showing the part number and a serial number on the driver's side. The serial number has no relationship to any other number. All chokes were conventional mechanical with a choke tube running from the unit to the driver's side exhaust manifold. The forward balance tube was coupled with an additional tube at the air cleaner adaptor and both ran across the front of the plenum. The forward balance tube goes to the fuel meter and the other tube to the oil filler. The cranking signal valve was attached directly to the enrichment diaphragm, and a line ran from the diaphragm to the plenum. Two lines were connected to the main diaphragm T, one going to the air meter, the other to the cranking signal valve. Distributor vacuum was supplied via a metal line from the air meter that wrapped around the rear of the plenum where it attached to a piece of rubber hose. Early units used a wobble pump (most of which were recalled) in the fuel meter. Usage was short lived because of mechanical problems and use of the gear pump resumed. Early units also had recesses under the nozzle blocks, cast bosses on the front of the plenum, and were without plenum flags.
7017375   360HP
    Air meter 7017376 is unstamped, (cast 7017248)
    Fuel meter 7017315 is unstamped, (cast 7017277)

FUEL PUMPS

The AC logo is cast directly in the top of the pump. The four digit Delco number is stamped in the edge of the mounting flange and should be 4657. Pumps are single diaphragm, screwed together assemblies, using short, 3/4" screws.

GAUGES

All gauges had silver concave cones and plastic lenses. A 60lb oil gauge was used until about mid production and was then changed to 80lb. There were two versions of the 7000 RPM tach face, 5300 red line for hydraulic lifter equipped cars, and 6500 red line for solid lifter cars. Early solid lifter versions had a warning buzzer that was activated when the tach reached the red area. Use of the buzzer was quickly discontinued, although the tachs continued to be installed until about January. Most of the buzzer tach faces had a cherry red warning area as opposed to the fluorescent orange found on the other models. All clocks were made by Borg Instruments and none of these were quartz, the second hand should show a definite ticking action, a continuous sweep indicates a quartz conversion. Many of the small gauges were serviced for quite some time, unfortunately they are not quite correct. The original faces were hot stamped leaving a slight depression where the characters are. Water soluble ink was also used so beware cleaning these with anything wet. The background black on the replacements is too flat and the characters are silk screened making them somewhat raised.

GLASS

All laminated glass was LOF Safety Plate and, along with the logo, was designated as such by AS1 for windshields and AS2 for side and rear glass. Side and rear Plexiglas was AS4. A two character code (month-year) also appeared indicating the date of manufacture. January=N, February=X, March=L, April=G, May=J, June=I, July=U, August=T, September=A, October=Y, November=C, December=V. The letter I indicates 1962, the letter C 1963.

GRILLE

Beware, some over the counter NOS grilles contain holes incorrectly drilled in the upper rail.

HARDTOP

All but very late production tops do not have reinforcement brackets attached with a button style retainer in the center of the rear window. The window is AS4 Plexiglas. Early production tops should not have the lower front edge notched for door clearance, as the later ones do. Headliners are made of a fibrous composition material, not vinyl. Headliner colors should match the interior.

HEADS

Casting # 3795896 for all 250HP cars, 3782461 for all others. The 3782461 heads contain a cast X on the combustion chamber side that is not part of the casting number. Dated month/day/year, with a single digit year representation. Double digits indicate Tonawanda manufacture and are not correct for Corvette. The month is indicated by a consecutive alphabetic character, ie January = A, February = B, etc. Symbol at the end of the head is a partial indication of the application.

HORNS

The last three digits of the part number are stamped into the horn. The high note is 9000456, the low note 9000455. The production date is also stamped year/month/week (1-5). Mounting brackets were welded to the horn, not bolted.

HORN RELAYS

1115824 was used on all cars. The cover is stamped with Delco Remy and the last three digits of the part number '824', along with '12V', are stamped under the mounting flange.

HUBCAPS

IGNITION SHIELDING

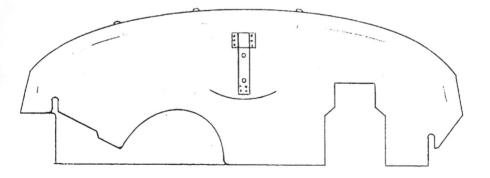

Early Clip
Til Mid '63

TOP
Three metal rivets hold the insulator. The bulge in the front skirt was higher than the later 1966-1967 top. The top did not have the CARB and FI slots in the rear skirt until the spring of 1963.

VERTICALS
The production right vertical was the same for 1963-1967. The left had a gradual curve to one side, at the upper 1/2 of the shield only. This curve runs from top to bottom in 1964-1967. The lower slanted edge is the same cut-away design used from 1964-1967, except for very early production.

HORIZONTALS
Two pieces of shielding were used on the driver's side. These were identical to the 1964-1965 pieces having the smaller exhaust manifold cut-out. A single long piece was used on the passenger side. This was identical to the 1962 part which had a reinforcement bar and large exhaust manifold cut-out.

INTAKE MANIFOLDS

Dated month/day/year. Casting numbers are located on the top surface. Cast iron manifold date codes are also located on the top surface, and are therefore visible with the manifold installed. Correct manifolds should have a single digit year representation. Aluminum manifold date codes are cast into the bottom surface and are not visible unless the manifold is removed from the engine. Cast iron 250HP manifold was 3783244. Cast iron 300HP manifold was 3799349. The 340HP aluminum manifold was 3794129.

JACK

The original equipment jack is very similar to those supplied for later model Corvettes, and even resembles some from other car models. The correct design has a stamped steel base, arms, and load rest, with a coarse thread jack screw. The load rest is angled to conform to the shape of the Corvette frame, and has the manufacturer's logo, 'A', stamped in it. The size of the base is 4"x6", and overall jack length is about 12 1/2", collapsed from arm to arm. Later jacks are larger. The trunions are the pieces that the jack screw pass through at each end of the jack. These should be U shaped, not solid, and the inside of the U faces the center of the jack with the nut on the outside. At the drive nut end, the load rest support arms are sandwiched between the trunion and the outer arms. These arms contain an oblong stamped reinforcement beginning mid production 1965. The drive nut end contains a single, large thrust washer, and the shaft is pinned at this end.

KNOBS

LAMPS

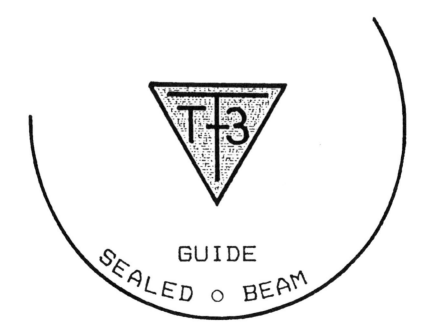

MIRRORS

Two styles of outside mirrors were used in 1963. The first was used until about April and had a long, thin base and pedestal with the ball stud placed in the lower 1/3 of the mirror back. The second style used only a stocky pedestal, also with the ball stud in the lower 1/3 of the mirror back. Both versions contained a bow tie emblem impression in the back of the mirror. A date code and manufacturers ID appeared on the glass in the form: month/ID/year

The inside mirrors were all basically the same with only minor variations. A mounting arm attaches to the upper windshield garnish molding and the mirror is attached to the arm with a screw. Early mirrors contained rivets on either side of the ball stud, some later mirrors did not. Later mirrors are dated with the same method as the outside mirrors, some early style mirrors were not dated.

OPTIONS

		Quantity	Price
30837	Base coupe	10,594	$4,252.00
30867	Base convertible	10,919	4,037.00
Incl.	327, 250HP	3,892	NC
Incl.	3 Speed	919	NC
Incl.	Black folding top	3,648	NC
A01	Tinted glass, all	629	16.15
A02	Tinted glass, windshield	470	10.80
A31	Power windows	3,742	59.20
C05	White folding top(instead of black)	5,728	NC
C05	Beige folding top(instead of black)	444	NC
C07	Hardtop	4,640	236.75
C07	Hardtop in place of soft top	1,099	NC
C48	Heater delete	124	-100.00
C60	Air conditioning	274	421.80
G81	Positraction rear	17,554	43.05
G91	Highway rear, 3.08 non posi	211	2.20
J50	Power brakes	3,336	43.05
J65	Metallic brakes	5,310	37.70
L75	327, 300HP engine	8,033	53.80
L76	327, 340HP engine	6,978	107.60
L84	327, 360HP engine	2,610	430.40
M20	Wide ratio 4 speed	8,444	188.30
M20	Close ratio 4 speed	9,529	188.30
M35	Powerglide	2,621	199.10
N03	36 Gallon fuel tank	63	202.30
N11	Off road exhaust	none delivered	
N34	Simulated wood wheel	130	16.15

1963 Corvette

Code	Option	Qty	Price
N40	Power steering	3,063	75.35
P48	Aluminum knock off wheels	none delivered	
P91	Nylon blackwalls	412	15.70
P92	Whitewalls	19,383	31.55
T86	Back up lamps	318	10.80
U65	Wonderbar AM radio	11,368	137.75
U69	AM/FM radio	9,178	174.35
Z06	Special performance equipment	199	1,818.45
Z12	Speedometer driven gear	for factory use	
898	Saddle leather seats	1,114	80.70
941	Sebring silver exterior paint		80.70

DEALER AVAILABLE OPTIONS

Option	Price
Fire Extinguisher	16.50
Portable spot lamp	8.50
Floor mats	9.25
Locking gas cap	5.75
Luggage carrier	32.50
Tool kit	6.25
Inside day-night mirror	12.20

PAINT/TRIM

Metal tag with paint and trim codes is located under the glove box.

EXTERIOR	CODE	INTERIOR
Tuxedo Black	900	Black:Red:Saddle
Silver Blue	912	Black:Blue
Daytona Blue	916	Red:Saddle:Blue
Riverside Red	923	Red:Black:Saddle
Saddle Tan	932	Black:Red:Saddle
Ermine White	936	Black:Red:Saddle:Blue
Sebring Silver	941	Black:Red:Saddle:Blue

These were factory recommended combinations.

A black, white, or beige convertible top was available with any paint/trim combination.

TRIM	COUPE/VINYL	CONV/VINYL	COUPE/LTHR	CONV/LTHR
Saddle	E,N,U,XJ,XL	F,P,V,XK,XM	A,E,Q,G,S	B,F,R,H,T
Blue	A,J,S,XE,XG	B,K,T,XF,XH	none	none
Red	C,L,Q,XA,XC	D,M,R,XB,XD	none	none
Black	STD,BLK	STD,BLK	none	none

With the exception of black, the number 490 preceded the code on all cars with vinyl trim, and 898 on all cars with leather trim.

PRODUCTION FIGURES	MONTHLY	CUMULATIVE
September	675	675
October	2081	2756
November	1291	4047
December	1925	5972
January	2004	7976
February	1838	9814
March	2019	11833
April	2295	14128
May	2281	16409
June	2115	18524
July	2466	20990
August	523	21513

RADIATORS

All radiators were aluminum, part number 3155316, with a separately mounted expansion tank. All radiators contained a stamp in the top that showed the part number and date of manufacture represented by a double digit year code and a single alphabetic month code (Jan=A, Feb=B, etc.).

RADIATOR CAPS

RADIOS

Model number 985396 Wonder Bar was offered at the beginning of the year, and were used until they were gone in the spring of 1963. This usage overlapped with the introduction and installation of the AM/FM radio, model number 985686.

Although the AM/FM was released for production in mid October installations did not begin until after the first of the year. The AM/FM dial face should have the AM indicator in red and the FM in green. A paper tag on the outside of one of the radio covers contains the model number.

REAR AXLE

Code and build date (month/day/year) are located on the rear, lower edge of the carrier.

CODES
CA .... 3.36, Open, all
CB .... 3.36, Positraction, all
CC .... 3.55, Positraction, 4 speed
CD .... 3.70, Positraction, 4 speed
CE .... 4.11, Positraction, 4 speed
CF .... 4.56, Positraction, 4 speed
CJ .... 3.08, Positraction, 4 speed
CX .... 3.70, Open, 4 speed
CZ .... 3.08, Open, 4 speed

SEATS

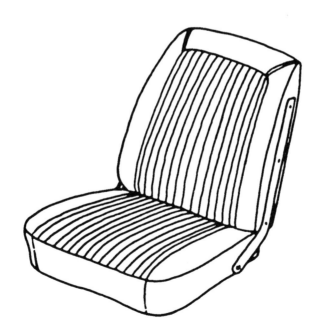

SHIFTERS

The three speed shifter and the automatic transmission shifter have a similar overall appearance, even including the mounting bracket. Both utilized a round, thin, chrome plated shift lever. Automatic shift pattern was P R N D L front to back.

The four speed shifter was also somewhat similar even with the addition of the reverse lockout. The shifter tower shaft contained a single spring for gate control tension. This was held in place by a round, slotted nut. An anti rattle spring was attached between the shift lever and a small bracket bolted to the bottom of the shifter support. The reverse lever should be about the same length as the third/fourth lever, unlike the earlier shifters where the reverse lever is distinctly longer.

SPARK PLUG WIRES

For carbureted cars the distributor boot is 180 degree, black. Spark plug boot is 90 degree, black. For fuel injected cars the distributor and spark plug boot is 90 degree, black. Wire is 7MM with the imprint PACKARD and an indication that the wire is radio/TV suppression, they are dated quarterly. The date appeared as 3-Q-62, for example, as the third quarter 1962.

STARTERS

1107242 was used on all cars. The date and model number were stamped directly into the starter housing. The date is shown as year, month, day, and the month appears as an alphabetic character with A=Jan, B=Feb, etc. The solenoid bakelite should contain Delco Remy.

STEERING WHEELS

Color matching (red, blue, tan, or black) wheels were plastic with metal spokes. Outer plastic circumference should have simulated grain finish. Until about July 1st, finishing marks on the spokes should run the same direction as the rim on the face of the spoke and not appear at all on the back. After July 1st the finishing marks are perpendicular to the rim. In mid March an optional simulated wood wheel was offered having all of the same characteristics of the colored wheel.

THERMOSTAT HOUSINGS

All special high performance solid lifter engines used an angled, cast iron housing, number 3827370. All other engines used a curved, cast iron housing, number 3827369.

TIRES

General 'Jet Air', Goodyear 'Power Cushion', Firestone 'Deluxe Champion', US Royal 'Safety 800', or B.F. Goodrich 'Silvertown' 6.70x15, tubeless. If whitewall equipped, the width should be 7/8" to 1 1/16" nominal. DOT (department of transportation) information did not appear in the sidewall of original tires.

TRANSMISSIONS

The four speed was available as an option in 1963 and two different units were used during the year. The first, a Borg Warner installed until about May, was unique to 1963. The main case was a T10D-1 which used a small front bearing retainer requiring a bellhousing with a correspondingly small support hole. The main case was cast aluminum and the cast date appeared within circles on the passenger side. The tail housing should have two shifter mounting positions (5 holes), and T10 7D should be cast along with the date. The side cover should contain webbing and a date on the outside. A build code is stamped on the upper rear, driver's side, unmachined part of the case. The code is in the form: location/month/day/year/shift, W=Warner Gear, month=Jan=A, Feb=B, etc., shift=1,2,3. Vehicle serial numbers are stamped on these transmissions next to the build information.

The Muncie four speed used in late production had the same unique small front bearing retainer as the Borg Warner described above. The tail housing is a 1731, and should have the speedometer gear on the driver's side. The side cover is a 707. Casting dates did not appear on these parts. A build code and the vehicle serial number is stamped on the rear of the case, driver's side. Build information is in the form: P(for Muncie)/month/day/shift (D or N).

Build information is stamped on the upper rear of the case, passenger side for three speeds. For Powerglide the stamping was on the passenger side of the oil pan. Build information is in the form: location/month/day/shift (D or N). S=Saginaw (three speed), T=Toledo (Powerglide). The three speed must have a shifter mounting position on the tail housing.

VALVE COVERS

All valve cover mounting holes were equally spaced at 8 3/4" top and bottom. The 250 and 300HP engine used a stamped steel cover with a rectangular raised area in the center. The raised area contained a label that read CHEVROLET at the top, 327 in the center, and TURBO-FIRE at the bottom. The sides of the raised area could be straight, (early) or triangular.

The remaining engines used a cast aluminum cover with CORVETTE in raised, 1/8" script. A casting flaw should not appear through the 'O' in CORVETTE. All covers had seven fins running the length of the cover. Two notches appeared on the inside lip for intake manifold clearance.

VOLTAGE REGULATORS

1119512 until about May, then mixed with 1119515. The cover was stamped with Delco Remy on two separate lines and was attached with slotted cap screws. The wire wound resistors underneath should both be the same physical size. The part number and date, along with '12VN' for 12 volt negative ground, are stamped in the mounting flange. The date is shown as year/month, with the month represented by an alphabetic character, ie A=Jan, B=Feb, etc. Beware of the reproductions with unequal size resistors and poor quality on the flange stamping.

WATER PUMPS

Casting # 3782608, for 250 & 300HP undated. Casting # 3859326, for 340 & 360HP, undated with a small hole to accept the 5/8"NP bypass fitting without using an adaptor.

WIPER ARMS & BLADES

All cars used a bright stainless and chrome arm with a bright stainless blade, Trico brand, with rubber insert. The Trico name may not always appear on the parts. The rubber inserts have a series of raised dots running the length of both sides. The blade style was a full length center hinge bridge, with two links almost hidden directly below the hinge. The blade top has a flat cross section.

WIPER MOTORS

5044518 stamped in the armature end housing. The washer pump nozzle should be one piece, translucent off-white.

1963

```
       JANUARY                    FEBRUARY
S  M  T  W  T  F  S       S  M  T  W  T  F  S
      1  2  3  4  5                       1  2
6  7  8  9 10 11 12       3  4  5  6  7  8  9
13 14 15 16 17 18 19     10 11 12 13 14 15 16
20 21 22 23 24 25 26     17 18 19 20 21 22 23
27 28 29 30 31           24 25 26 27 28

        MARCH                      APRIL
S  M  T  W  T  F  S       S  M  T  W  T  F  S
                1  2          1  2  3  4  5  6
3  4  5  6  7  8  9       7  8  9 10 11 12 13
10 11 12 13 14 15 16     14 15 16 17 18 19 20
17 18 19 20 21 22 23     21 22 23 24 25 26 27
24 25 26 27 28 29 30     28 29 30
31

         MAY                       JUNE
S  M  T  W  T  F  S       S  M  T  W  T  F  S
         1  2  3  4                           1
5  6  7  8  9 10 11       2  3  4  5  6  7  8
12 13 14 15 16 17 18      9 10 11 12 13 14 15
19 20 21 22 23 24 25     16 17 18 19 20 21 22
26 27 28 29 30 31        23 24 25 26 27 28 29
                         30

        JULY                      AUGUST
S  M  T  W  T  F  S       S  M  T  W  T  F  S
   1  2  3  4  5  6                   1  2  3
7  8  9 10 11 12 13       4  5  6  7  8  9 10
14 15 16 17 18 19 20     11 12 13 14 15 16 17
21 22 23 24 25 26 27     18 19 20 21 22 23 24
28 29 30 31              25 26 27 28 29 30 31

       SEPTEMBER                  OCTOBER
S  M  T  W  T  F  S       S  M  T  W  T  F  S
1  2  3  4  5  6  7             1  2  3  4  5
8  9 10 11 12 13 14       6  7  8  9 10 11 12
15 16 17 18 19 20 21     13 14 15 16 17 18 19
22 23 24 25 26 27 28     20 21 22 23 24 25 26
29 30                    27 28 29 30 31

       NOVEMBER                  DECEMBER
S  M  T  W  T  F  S       S  M  T  W  T  F  S
                1  2       1  2  3  4  5  6  7
3  4  5  6  7  8  9       8  9 10 11 12 13 14
10 11 12 13 14 15 16     15 16 17 18 19 20 21
17 18 19 20 21 22 23     22 23 24 25 26 27 28
24 25 26 27 28 29 30     29 30 31
```

1964

```
       JANUARY                    FEBRUARY
S  M  T  W  T  F  S       S  M  T  W  T  F  S
         1  2  3  4                           1
5  6  7  8  9 10 11       2  3  4  5  6  7  8
12 13 14 15 16 17 18      9 10 11 12 13 14 15
19 20 21 22 23 24 25     16 17 18 19 20 21 22
26 27 28 29 30 31        23 24 25 26 27 28 29

        MARCH                      APRIL
S  M  T  W  T  F  S       S  M  T  W  T  F  S
1  2  3  4  5  6  7                1  2  3  4
8  9 10 11 12 13 14       5  6  7  8  9 10 11
15 16 17 18 19 20 21     12 13 14 15 16 17 18
22 23 24 25 26 27 28     19 20 21 22 23 24 25
29 30 31                 26 27 28 29 30

         MAY                       JUNE
S  M  T  W  T  F  S       S  M  T  W  T  F  S
               1  2       1  2  3  4  5  6
3  4  5  6  7  8  9       7  8  9 10 11 12 13
10 11 12 13 14 15 16     14 15 16 17 18 19 20
17 18 19 20 21 22 23     21 22 23 24 25 26 27
24 25 26 27 28 29 30     28 29 30
31

        JULY                      AUGUST
S  M  T  W  T  F  S       S  M  T  W  T  F  S
         1  2  3  4                            1
5  6  7  8  9 10 11       2  3  4  5  6  7  8
12 13 14 15 16 17 18      9 10 11 12 13 14 15
19 20 21 22 23 24 25     16 17 18 19 20 21 22
26 27 28 29 30 31        23 24 25 26 27 28 29
                         30 31

       SEPTEMBER                  OCTOBER
S  M  T  W  T  F  S       S  M  T  W  T  F  S
      1  2  3  4  5                   1  2  3
6  7  8  9 10 11 12       4  5  6  7  8  9 10
13 14 15 16 17 18 19     11 12 13 14 15 16 17
20 21 22 23 24 25 26     18 19 20 21 22 23 24
27 28 29 30              25 26 27 28 29 30 31

       NOVEMBER                  DECEMBER
S  M  T  W  T  F  S       S  M  T  W  T  F  S
1  2  3  4  5  6  7             1  2  3  4  5
8  9 10 11 12 13 14       6  7  8  9 10 11 12
15 16 17 18 19 20 21     13 14 15 16 17 18 19
22 23 24 25 26 27 28     20 21 22 23 24 25 26
29 30                    27 28 29 30 31
```

1964 Vehicle Identification Number (VIN) 408x7S100001 to 408x7S122229, total vehicles 22,229, 13,925 convertibles, 8,304 coupes. Plates are located under the glove compartment, spot welded to the horizontal brace.
First digit=model year
Fourth digit=6 for convertible, 3 for coupe
S=St. Louis
Six digit sequential serial number starting at 100001

AIR CLEANERS

The 250 and 300HP engines used a dual snorkel style air cleaner. The lid was chrome plated and removable to allow servicing of the original foam element. The base was painted black and had a large, approximately 1", tube extending from the rear to facilitate the hose connection to the breather tube. Cars with air conditioning had pieces attached to the carburetor cutout in order to shift the snorkels away from the radiator hose. The 365HP air cleaner was basically the same except the base was chrome plated, had a small warm air tube connection, and had cutouts in the rear for air intake replacing the snorkels in the front.

The fuel injection air cleaner, and S tube connection to the core support were originally metal. The main canister was mounted to the driver's side inner fender. An elbow was attached to the main canister with three latch type clamps allowing access to the filter element. The uppermost of these three clamps was mounted on top of, not in front of, the fender mounting flange.

AIR CONDITIONING

Extensive changes were required, both small and large, to accommodate the additional equipment necessary for this option. Major additional items included the condenser, mounted in front of the radiator; the evaporator and case, mounted on the passenger side firewall; the dehydrator (drier), mounted on the core support; the compressor, mounted on the engine, passenger side; two additional dash controls; interior ducting consisting of a small outlet above the clock, and two large outlets mounted under the dash at both ends. Adding these parts required different passenger side horn positioning, a new windshield washer reservoir (a bag instead of a plastic jar), battery relocation to the driver's side, moving the headlight motor switch, a longer hood release bracket. and a different trip odometer cable bracket, among other obvious things such as wiring, etc.

The parts themselves evolved into different versions depending on design and manufacturing changes. The compressor mounting was accomplished using a different passenger side exhaust manifold from non air conditioned cars, and the brackets were all cast. The compressor itself had a Frigidaire identification tag on the case with model and serial number, and the rear of the case was dated month/day/year. The month was represented by a letter, Jan=A, Feb=B, etc.

The suction throttling valve, or STV was used to control evaporator pressure from 1963 to 1966, and is mounted on the evaporator case. Both large lines should enter at the top, not top and bottom as seen on POA valves used as replacements.

Original water control valves have a hexagonal portion as part of their construction that was not retained by the later replacements.

Dehydrators (driers) were originally manufactured with a single continuous diameter, later replacements look like an old milk bottle with a smaller diameter at the top.

Replacement or reproduction condensers are more of a universal type having mounting flanges that, unlike the originals, accommodate a number of applications.

The center and right side ducts were the same from 1964 thru 1967. The left side attaches to the side of the steering column brace from 1964 thru 1966. The 1967 left side duct attaches to the underside of the dash brace.

ALTERNATORS

1100668/37 Amp, all except cars equipped with air conditioning and/or transistor ignition
1100665/55 Amp, with air conditioning
1100669/42 Amp, with transistor ignition
1100684/60 Amp, with air conditioning and transistor ignition

All units are stamped with the amperage rating, the model number, and the date. The date is represented year/month/day, with a letter code used for the month, ie A=Jan, B=Feb, etc. Stampings were on the front case. Four blades on the fan should be wider than the others, and the fan should be painted black. The pulley should be cad plated. Cars with solid lifters without air conditioning used a wide, 3 5/8" diameter pulley with a 1/2" belt groove, with air conditioning the diameter was 3 3/64". Cars with hydraulic lifters used a narrow 2 13/16" diameter pulley with a 3/8" belt groove.

ANTENNAS

Assembly 3832327. Three section mast without rings on any section, ball at the top of the top section. Base consists of a round chrome nut with notches for a spanner wrench, a chrome bezel, and a gasket. The fact that the top two sections must collapse completely into the bottom section in order for the antenna to be authentic is erroneous.

BALANCERS

Narrow, 6" diameter balancer with the pulley bolted directly to it was used on all engines with hydraulic lifters. The solid lifter motors used a 1 11/16" wide, 8" diameter balancer with a bolt on pulley. This balancer was dated month/year on the rear face and also contained cooling fins projecting from the inner rear flange.

BATTERIES

1980558 tar top. Year/month and factory of origin is stamped in the top. Cover above the tar on top should be 2 inches wide. Caps are all black with Delco in yellow on the outside, visible casting line/flashing on the bottom, inside. Case side should say "Delco Original Equipment Power Rated" and should be unpainted (warranty replacements were painted). The characters 559 should be stamped in the side and CAT 558 on top of the battery.

BATTERY CABLES

Both cables are spring style and vinyl covered. The part number is stamped near the terminal end. The negative terminal contains an N and the positive a P, but the P will be covered by a short black sleeve. The positive cable is black the negative brown.

BELLHOUSINGS

3858403 aluminum, dated month-day in a small circle usually unreadable.

BRAKE MASTER CYLINDERS

Except for the special J56 brake package, the 7/8" bore size identification is cast into the top of the master cylinder at the front. The cap is held in place by a wire bail and contains USE DELCO BRAKE FLUID stamped equidistant around the dome. The J56 master cylinder was a dual line unit with a screw on cap.

CARBURETORS

Except for Holley, model numbers are embossed on an aluminum tag attached under one of the carburetor top screws. The date is also shown on this tag in the form month/year, using A=Jan, B=Feb, etc., an additional digit indicated factory production information. AFB carburetors have the model number and date stamped on the front of the base, passenger side. The date is in the same form as the aluminum tag.

Holley information was stamped into the front of the air horn. At the top was the Chevrolet part number, the second line was the LIST, or Holley model number, the third line was the date. The date is in the form, year/month/week, each represented by one character. The characters 1 thru 0 represent January thru October, with A and B representing November and December respectively. Be aware that four digit date codes were started later (1970's) in production and are not correct.

All 250 HP with manual transmission used WCFB 3697S, with Powerglide WCFB 3696S. They have the choke mounted at the bottom, and the choke line is attached to the center of the choke cover. A secondary counterweight is used. The main body is cast 0-1465, top 6-1672.

The 300 HP with Powerglide used AFB 3720S, SA or SB. The 300 HP manual transmission cars used AFB 3721S, SA, or SB. The main body is cast 0-1552, top 6-1518 for both.

The 365 HP used a Holley R2818A. The date code should be correct, and there should be visible float adjusting screws and a secondary metering block.

CARPET

Loop carpet, foam backed, and sectionally sewn to fit the individual sections and contours of the interior.

COILS

1115087, hydraulic lifter equipped cars. 1115091, solid lifter equipped cars without transistor ignition. 1115176, cars with transistor ignition. The last 3 digits of the part number appear in raised characters on the case.

CONVERTIBLE TOP

All tops had a rear window attached with heat sealing from the inside. The rear window contained the Vinylite name and trademark, the AS 6 identification, cleaning instructions, and a date code hot stamped into the outside surface. The top frame differed in one respect from 1963 to 1967. The first bow, directly above the driver, was round from 1963 to mid 1965. The pads attached to this bow with screws. The bow used from mid 1965 to 1967 was flat, and the pads were attached to a tack strip.

CRANKCASE VENTS

The rear of the engine block contains a large pipe assembly that is connected to the air cleaner base or FI adaptor with a short piece of rubber hose. A flame arrestor was inserted in this hose on fuel injected cars. The support bracket for the pipe is attached to the outside of the intake manifold on 250 and 300HP, and towards the center of the intake manifold on 365HP. The pipe on fuel injected engines did not have a support bracket. The oil filler tube had a small pipe that was connected via a piece of rubber hose to the carburetor or fuel injection plenum chamber. Metered orifices were used as opposed to PCV valves. Early fuel injected cars may have used 1963 components.

DISTRIBUTORS

The models used with the carbureted engines contain a metal tag that wraps around the neck of the distributor below the cap. This contains the date and model number. All 250, and 300HP cars used model number 1111024. The early 365HP without transistor ignition used 1111062, late used 1111069. Cars with 365HP and transistor ignition used 1111060. The fuel injection cars without transistor ignition used model 1111063, those with transistor ignition used 1111064. Fuel injection models had a cross shaft access cover. The date and model number were stamped in the cover. All models were dated year/month/day. All models also contained a facility for driving the tachometer cable.

EMBLEMS

Rear Deck

Nose

Side

Side

ENGINE BLOCKS

Casting # 3782870 in characters 1/2" high. Smaller characters were used on passenger car blocks. All engines were 327 cubic inches. Casting number is located on the driver's side rear flange where the bellhousing attaches. The date is located directly opposite on the passenger side rear flange. They are dated month/day/year, with a single digit year representation. Double digits indicate Tonawanda manufacture and are not correct for Corvette. The month is indicated by a consecutive alphabetic character, ie January = A, February = B, etc. A code indicating the assembly location, date, and engine characteristics was stamped on a pad located on the passenger side of the motor. It is just below the cylinder head, next to the water pump. The first character should be F for the Flint assembly plant. The numeric characters represent the date, and the last two characters indicate the engine type as shown in the chart that follows. The pad also contained a vehicle serial number, separate from, and to the left of the assembly code.

RC = STD 250HP, 4 barrel, manual transmission
SC = STD 250HP, 4 barrel, automatic transmission
RP = STD 250HP, 4 barrel, manual, air conditioning
SK = STD 250HP, 4 barrel, automatic, air conditioning
RD = L75 300HP, 4 barrel, high performance, manual
SD = L75 300HP, 4 barrel, high performance, automatic
RQ = L75 300HP, 4 barrel, high performance, manual, air conditioning
SL = L75 300HP, 4 barrel, high performance, automatic, air conditioning
RE = L76 365HP, 4 barrel, special high performance, manual transmission
RR = L76 365HP, 4 barrel, special high performance, air conditioning, manual transmission
RT = L76 365HP, 4 barrel, special high performance, transistor ignition, manual transmission
RU = L76 365HP, 4 barrel, special high performance, transistor ignition, air conditioning, manual transmission
RF = L84 375HP, fuel injection, manual transmission
RX = L84 375HP, fuel injection, transistor ignition, manual transmission

EXHAUST MANIFOLDS

All 250HP, and 300HP with automatic transmission 3846559 left, 3750556 right, or 3747038 right with air conditioning, dated month-day only. The 300HP with manual transmission and 365HP used 3846563 left, 3797902 right, or 3797942 right, with air, all undated. The 375HP used 3797942 left, 3797902 right, both undated.

FUEL FILTERS

Cars with a WCFB four barrel used a filter mounted in the carburetor just behind the inlet fitting. All other cars used an external GF 90 filter. The color of the GF 90 was black.

FUEL INJECTION

Two units were used in 1964 and they used a wide, removable, ribbed plenum cover. The plenum contained a black/aluminum tag showing the part number and a serial number on the driver's side. This serial number has no relationship to any other number. GM began stamping the vehicle serial number on the plenum early in 1964. All chokes were conventional mechanical with a choke tube running from the unit to the driver's side exhaust manifold.

7017375R 375HP. Early production only. The forward balance tube was coupled with an additional tube at the air cleaner adaptor and both ran across the front of the plenum. The forward balance tube goes to the fuel meter and the other tube to the oil filler. The cranking signal valve was attached directly to the enrichment diaphragm, and a line ran from the diaphragm to the plenum. Two lines were connected to the main diaphragm T, one going to the air meter, the other to the cranking signal valve. Distributor vacuum was supplied via a metal line from the air meter that wrapped around the rear of the plenum where it attached to a piece of rubber hose. Actually a recalibrated 1963 unit.
Air meter 7017376 unstamped, (cast 7017248)
Fuel meter 7017315 unstamped, (cast 7017277)

7017380 375HP. A solenoid activated by a micro switch replaced the cranking signal valve for starting purposes. Distributor vacuum was taken directly from the rear side of the plenum. The oil filler tube connected to a restrictor fitting installed in the front of the plenum. One line ran from the main diaphragm to the air meter. One line also ran from the enrichment diaphragm to the plenum. The forward balance tube ran from the fuel meter to the air cleaner adaptor. Late production had a splash shield attaching bolt located on the rear of the main diaphragm cover that was locked in place with a nut.
Air meter 7029539 unstamped, (cast 7017248)
Fuel meter 7017377 unstamped, (cast 7017277)

FUEL PUMPS

The AC logo is cast directly in the top of the pump. A four digit Delco number is stamped in the edge of the mounting flange and should be 4657 for cars with hydraulic lifters, and without air conditioning. A five digit number, 40083, is stamped in the flange of solid lifter cars and air conditioned cars. Pumps are single diaphragm, screwed together assemblies, using short, 3/4" screws.

GAUGES

All gauges had black concave cones and glass lenses with the exception of the clock which had a plastic lens. An 80lb oil gauge was used for the entire year regardless of the motor. There were two versions of the 7000 RPM tach face, 5300 red line for hydraulic lifter equipped cars, and 6500 red line for solid lifter cars. All clocks were made by Borg Instruments and none of these were quartz, the second hand should show a definite ticking action, a continuous sweep indicates a quartz conversion. Many of the small gauges were serviced for quite some time, unfortunately they are not quite correct. The original faces were hot stamped leaving a slight depression where the characters are. Water soluble ink was also used so beware cleaning these with anything wet. The background black on the replacements is too flat and the characters are silk screened making them somewhat raised.

GLASS

All laminated glass was LOF Safety Plate and, along with the logo, was designated as such by AS1 for windshields and AS2 for side and rear glass. Side and rear Plexiglas was AS4. A two character code (month-year) also appeared indicating the date of manufacture. January=N, February=X, March=L, April=G, May=J, June=I, July=U, August=T, September=A, October=Y, November=C, December=V. The letter C indicates 1963, the letter G 1964.

GRILLE

HARDTOP

All tops have reinforcement brackets attached with a button style retainer in the center of the rear window. The window is AS4 Plexiglas. Headliners are made of a fibrous composition material, not vinyl. Headliner colors should match the interior.

HEADS

Casting #3795896 for all 250HP cars, 3782461 for all others. Dated month/day/year, with a single digit year representation. Double digits indicate Tonawanda manufacture and are not correct for Corvette. The month is indicated by a consecutive alphabetic character, ie January = A, February = B, etc. Symbol at the end of the head is a partial indication of the application.

HORNS

The first four digits of the part number are cast into the horn followed by three stamped digits. The high note is 9000488, the low note 9000487. The horns are also production date stamped year/month/week (1-5). Passenger car horns are assembled with the trumpet indexed at a different location, and the mounting bracket is also different. Mounting brackets were welded to the horn, not bolted.

HORN RELAYS

1115824 was used on all cars. The cover is stamped with Delco Remy and the last three digits of the part number '824', along with '12V', are stamped under the mounting flange.

HUBCAPS

IGNITION SHIELDING

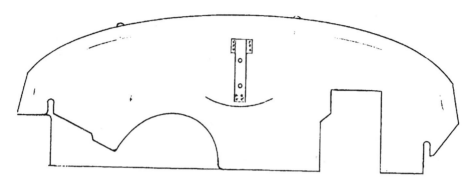

TOP

Two metal rivets hold the insulator. The bulge in the front skirt was higher than the later 1966-1967 top.

VERTICALS

The production right vertical was the same for 1963-1967. The left had a gradual curve to one side, running from top to bottom from 1964-1967.

HORIZONTALS

Two pieces of shielding on each side. All 1964-1965 pieces have the smaller exhaust manifold cut-out.

INTAKE MANIFOLDS

Dated month/day/year. Casting numbers are located on the top surface. Cast iron manifold date codes are also located on the top surface, and are therefore visible with the manifold installed. Correct manifolds should have a single digit year representation. Aluminum manifold date codes are cast into the bottom surface and are not visible unless the manifold is removed from the engine. Cast iron 250HP manifold was 3844457. Cast iron 300HP manifold was 3844459. The 365HP aluminum manifold was 3844461.

JACK

The original equipment jack is very similar to those supplied for later model Corvettes, and even resembles some from other car models. The correct design has a stamped steel base, arms, and load rest, with a coarse thread jack screw. The load rest is angled to conform to the shape of the Corvette frame, and has the manufacturer's logo, 'A', stamped in it. The size of the base is 4"x6", and overall jack length is about 12 1/2", collapsed from arm to arm. Later jacks are larger. The trunions are the pieces that the jack screw pass through at each end of the jack. These should be U shaped, not solid, and the inside of the U faces the center of the jack with the nut on the outside. At the drive nut end, the load rest support arms are sandwiched between the trunion and the outer arms. These arms contain an oblong stamped reinforcement beginning mid production 1965. The drive nut end contains a single, large thrust washer, and the shaft is pinned at this end.

KNOBS

LAMPS

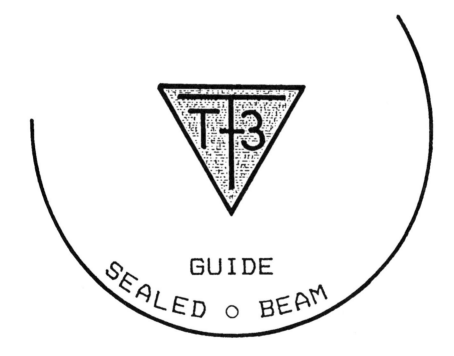

MIRRORS

The outside mirror used a stocky pedestal with the ball stud in the lower 1/3 of the mirror back. A bow tie emblem impression appeared in the back of the mirror. A date code and manufacturers ID appeared on the glass in the form: month/ID/year.

The inside mirrors were all basically the same with only minor variations. A mounting arm attaches to the upper windshield garnish molding and the mirror is attached to the arm with a screw. Some mirrors contained rivets on either side of the ball stud, some mirrors did not. Inside mirrors are dated with the same method as the outside mirrors.

OPTIONS

Code	Description	Quantity	Price
40837	Base coupe	8,304	$4,252.00
40867	Base convertible	13,925	4,037.00
Incl.	327, 250HP	3,262	NC
Incl.	3 Speed	715	NC
Incl.	Black folding top	4,721	NC
A01	Tinted glass, all	6,031	16.15
A02	Tinted glass, windshield	6,387	10.80
A31	Power windows	3,706	59.20
C05	White folding top(instead of black)	4,843	NC
C05	Beige folding top(instead of black)	591	NC
C07	Hardtop	5,803	236.75
C07	Hardtop in place of soft top	1,220	NC
C48	Heater delete	60	-100.00
C60	Air conditioning	1,988	421.80
F40	HD Suspension	82	37.70
G81	Positraction rear	18,279	43.05
G91	Highway rear, 3.08 non posi	2,310	2.20
J50	Power brakes	2,270	43.05
J56	Special brakes	29	629.50
J65	Metallic brakes	4,780	53.80
K66	Transistor ignition	552	75.35
L75	327, 300HP engine	10,471	53.80
L76	327, 365HP engine	7,171	107.60
L84	327, 375HP engine	1,325	538.00
M20	Wide ratio 4 speed	10,538	188.30
M20	Close ratio 4 speed	8,496	188.30
M35	Powerglide	2,480	199.10
N03	36 Gallon fuel tank	38	202.30
N11	Off road exhaust	1,953	37.70
N40	Power steering	3,126	75.35
P48	Aluminum knock off wheels	806	322.80
P91	Nylon blackwalls	372	15.70
P92	Whitewalls	19,977	31.85
T86	Back up lamps	11,085	10.80

U69	AM/FM radio	20,934	176.50
Z12	Speedometer driven gear	for factory use	
-	Leather seats	1,334	80.70

DEALER AVAILABLE OPTIONS

Fire Extinguisher	14.95
Portable spot lamp	8.50
Floor mats	9.25
Locking gas cap	5.75
Luggage carrier	32.50
Tool kit	6.50
Inside day-night mirror	12.20

PAINT/TRIM

Metal tag with paint and trim codes is located under the glove box.

EXTERIOR	CODE	INTERIOR
Tuxedo Black	900	Black:Red:Silver:White
Silver Blue	912	Black:Blue:White
Daytona Blue	916	Blue:Silver:White
Riverside Red	923	Red:Black:White
Saddle Tan	932	Saddle:White
Ermine White	936	Black:Red:Saddle:Blue:Silver:White
Satin Silver	940	Black:Red:Blue:Silver:White

These were factory recommended combinations.

A black, white, or beige convertible top was available with any paint/trim combination.

TRIM	COUPE/VINYL	CONV/VINYL	COUPE/LTHR	CONV/LTHR
Saddle	490CA,L	490CB,M	898CA,G	898DA,H
Blue	490BA,J	490BB,K	898JA,N	898KA,P
Red	490AA,G	490AB,H	898EA,L	898FA,M
Black	STD,BLK	STD,BLK	898A	898A
Sil/Blk	491AA	491AE	899AA	899AE
Sil/Blu	491BA,M	491BE,N	899BA,M	899BE,N
Wht/Blk	491CA	491CE,CB	899CA	899CE,CB
Wht/Blu	491GA,R	491GE,S	899GA,R	899GE,S
Wht/Red	491DA,P	491DE,Q	899DA,P	899DE,Q
Wht/Sad	491HA,T	491HE,U	899HA,T	899HE,U

Silver and black used a black dash and gray carpet.
Silver and blue used a blue dash and gray carpet.
White and black used a black dash and carpet.
White and blue used a blue dash and carpet.
White and red used a red dash and carpet.
White and saddle used a saddle dash and carpet.

PRODUCTION FIGURES

	MONTHLY	CUMULATIVE
September	1741	1741
October	2304	4045
November	2018	6063
December	2028	8091
January	2206	10297
February	2025	12322
March	2248	14570
April	2295	16865
May	1940	18805
June	2115	20920
July	1309	22229

RADIATORS

All radiators were aluminum, part number 3155316, with a separately mounted expansion tank. All radiators contained a stamp in the top that showed the part number and date of manufacture represented by a double digit year code and a single alphabetic month code (Jan=A, Feb=B, etc.).

RADIATOR CAPS

With A/C

RADIOS

Model number 985921, AM/FM. The dial face shows AM in red and FM in green. A paper tag on the outside of one of the radio covers contains the model number.

REAR AXLE

Code and build date (month/day/year) are located on the rear, lower edge of the carrier.

```
CODES
CA .......... 3.36, Open, all
CB .......... 3.36, Positraction, all
CC .......... 3.55, Positraction, 4 speed
CD .......... 3.70, Positraction, 4 speed
CE .......... 4.11, Positraction, 4 speed
CF .......... 4.56, Positraction, 4 speed
CJ ........... 3.08, Positraction, 4 speed
CX .......... 3.70, Open, 4 speed
CZ .......... 3.08, Open, 4 speed
```

SEATS

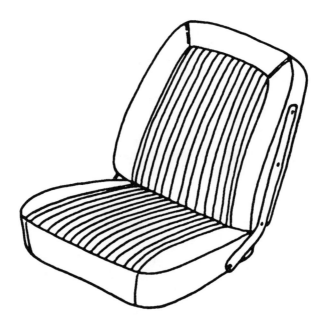

SHIFTERS

The automatic transmission shifter used the earlier thin shift lever, about 3/8" diameter. The automatic shift pattern remained P R N D L front to back. The three and four speed shifters used the new 3/4" diameter lever. The four speed lever had a reverse lockout and should not be confused with the later shifter that had a 1/2" longer lever. Lever length should be 6 3/8" from the base to the top without the ball. The later lever also has a greater side bend at the base.

SPARK PLUG WIRES

For carbureted cars the distributor boot is 180 degree, black. Spark plug boot is 90 degree, black. For fuel injected cars the distributor and spark plug boot is 90 degree, black. Wire is 7MM with the imprint PACKARD and an indication that the wire is radio/TV suppression, they are dated quarterly. The date appeared as 3-Q-62, for example, as the third quarter 1962.

STARTERS

1107320 was used on all cars. The date and model number were stamped directly into the starter housing. The date is shown as year, month, day, and the month appears as an alphabetic character with A=Jan, B=Feb, etc. The solenoid bakelite should contain Delco Remy.

STEERING WHEELS

All wheels were plastic simulated wood with metal spokes. The three spokes and center area hub attachment are one piece construction as opposed to the 1967 style which has the center spoke mounted in back of the others.

THERMOSTAT HOUSINGS

All special high performance solid lifter engines used an angled, cast iron housing, number 3827370. All other engines used a curved, cast iron housing, number 3827369.

TIRES

General 'Jet Air', Goodyear 'Power Cushion', Firestone 'Deluxe Champion', US Royal 'Safety 800', or B.F. Goodrich 'Silvertown' 6.70x15, tubeless. If whitewall equipped, the width should be 7/8" to 1 1/16" nominal. DOT (department of transportation) information did not appear in the sidewall of original tires.

TRANSMISSIONS

The optional four speed was a Muncie for the entire production year. The front bearing retainer was the larger style as opposed to the small 1963 type. The tail housing is a 429, and should have the speedometer gear on the driver's side. The side cover is a 707. Casting dates did not appear on these parts. A build code and the vehicle serial number is stamped on the rear of the case, driver's side.

Build information is stamped on the upper rear of the case, passenger side for three speeds. For Powerglide the stamping was on the passenger side of the oil pan. Build information for all transmissions is in the form: location/month/day/shift (D or N). S=Saginaw (three speed), T=Toledo (Powerglide), P=Muncie (four speed). The three speed must have a shifter mounting position on the tail housing.

VALVE COVERS

All valve cover mounting holes were equally spaced at 8 3/4" top and bottom. The 250 and 300HP engine used a stamped steel cover with a rectangular raised area in the center. The raised area contained a label that read CHEVROLET at the top, 327 in the center, and TURBO-FIRE at the bottom. The sides of the raised area should be triangular.

The remaining engines used a cast aluminum cover with CORVETTE in raised, 1/8" script. A casting flaw should not appear through the 'O' in CORVETTE. All covers had seven fins running the length of the cover. Two notches appeared on the inside lip for intake manifold clearance.

VOLTAGE REGULATORS

1119515 used on all cars. The cover was stamped with Delco Remy on two separate lines and was attached with slotted cap screws. The wire wound resistors underneath should both be the same physical size. The part number and date, along with '12VN' for 12 volt negative ground, are stamped in the mounting flange. The date is shown as year/month, with the month represented by an alphabetic character, ie A=Jan, B=Feb, etc. Beware of the reproductions with unequal size resistors and poor quality on the flange stamping.

WATER PUMPS

Casting # 3782608, for 250 & 300HP undated. Casting # 3859326, for 365 & 375HP, undated with a small hole to accept the 5/8"NP bypass fitting without using an adaptor.

WIPER ARMS & BLADES

All cars used a bright stainless and chrome arm with a bright stainless blade, Trico brand, with rubber insert. The Trico name may not always appear on the parts. The rubber inserts have a series of raised dots running the length of both sides. The blade style was a full length center hinge bridge, with two links almost hidden directly below the hinge. The blade top has a flat cross section.

WIPER MOTORS

5044518 stamped in the armature end housing. The washer pump nozzle should be one piece, translucent off-white.

1964

```
        JANUARY                    FEBRUARY
S  M  T  W  T  F  S       S  M  T  W  T  F  S
         1  2  3  4                         1
5  6  7  8  9 10 11       2  3  4  5  6  7  8
12 13 14 15 16 17 18      9 10 11 12 13 14 15
19 20 21 22 23 24 25     16 17 18 19 20 21 22
26 27 28 29 30 31        23 24 25 26 27 28 29

         MARCH                       APRIL
S  M  T  W  T  F  S       S  M  T  W  T  F  S
1  2  3  4  5  6  7                1  2  3  4
8  9 10 11 12 13 14       5  6  7  8  9 10 11
15 16 17 18 19 20 21     12 13 14 15 16 17 18
22 23 24 25 26 27 28     19 20 21 22 23 24 25
29 30 31                 26 27 28 29 30

          MAY                        JUNE
S  M  T  W  T  F  S       S  M  T  W  T  F  S
               1  2          1  2  3  4  5  6
3  4  5  6  7  8  9       7  8  9 10 11 12 13
10 11 12 13 14 15 16     14 15 16 17 18 19 20
17 18 19 20 21 22 23     21 22 23 24 25 26 27
24 25 26 27 28 29 30     28 29 30
31

         JULY                       AUGUST
S  M  T  W  T  F  S       S  M  T  W  T  F  S
         1  2  3  4                         1
5  6  7  8  9 10 11       2  3  4  5  6  7  8
12 13 14 15 16 17 18      9 10 11 12 13 14 15
19 20 21 22 23 24 25     16 17 18 19 20 21 22
26 27 28 29 30 31        23 24 25 26 27 28 29
                         30 31

       SEPTEMBER                   OCTOBER
S  M  T  W  T  F  S       S  M  T  W  T  F  S
      1  2  3  4  5                   1  2  3
6  7  8  9 10 11 12       4  5  6  7  8  9 10
13 14 15 16 17 18 19     11 12 13 14 15 16 17
20 21 22 23 24 25 26     18 19 20 21 22 23 24
27 28 29 30              25 26 27 28 29 30 31

       NOVEMBER                   DECEMBER
S  M  T  W  T  F  S       S  M  T  W  T  F  S
1  2  3  4  5  6  7          1  2  3  4  5
8  9 10 11 12 13 14       6  7  8  9 10 11 12
15 16 17 18 19 20 21     13 14 15 16 17 18 19
22 23 24 25 26 27 28     20 21 22 23 24 25 26
29 30                    27 28 29 30 31
```

1965

```
        JANUARY                    FEBRUARY
S  M  T  W  T  F  S       S  M  T  W  T  F  S
               1  2       1  2  3  4  5  6
3  4  5  6  7  8  9       7  8  9 10 11 12 13
10 11 12 13 14 15 16     14 15 16 17 18 19 20
17 18 19 20 21 22 23     21 22 23 24 25 26 27
24 25 26 27 28 29 30     28
31

         MARCH                       APRIL
S  M  T  W  T  F  S       S  M  T  W  T  F  S
   1  2  3  4  5  6                   1  2  3
7  8  9 10 11 12 13       4  5  6  7  8  9 10
14 15 16 17 18 19 20     11 12 13 14 15 16 17
21 22 23 24 25 26 27     18 19 20 21 22 23 24
28 29 30 31              25 26 27 28 29 30

          MAY                        JUNE
S  M  T  W  T  F  S       S  M  T  W  T  F  S
                  1          1  2  3  4  5
2  3  4  5  6  7  8       6  7  8  9 10 11 12
9 10 11 12 13 14 15      13 14 15 16 17 18 19
16 17 18 19 20 21 22     20 21 22 23 24 25 26
23 24 25 26 27 28 29     27 28 29 30
30 31

         JULY                       AUGUST
S  M  T  W  T  F  S       S  M  T  W  T  F  S
            1  2  3       1  2  3  4  5  6  7
4  5  6  7  8  9 10       8  9 10 11 12 13 14
11 12 13 14 15 16 17     15 16 17 18 19 20 21
18 19 20 21 22 23 24     22 23 24 25 26 27 28
25 26 27 28 29 30 31     29 30 31

       SEPTEMBER                   OCTOBER
S  M  T  W  T  F  S       S  M  T  W  T  F  S
      1  2  3  4                         1  2
5  6  7  8  9 10 11       3  4  5  6  7  8  9
12 13 14 15 16 17 18     10 11 12 13 14 15 16
19 20 21 22 23 24 25     17 18 19 20 21 22 23
26 27 28 29 30           24 25 26 27 28 29 30
                         31

       NOVEMBER                   DECEMBER
S  M  T  W  T  F  S       S  M  T  W  T  F  S
   1  2  3  4  5  6                1  2  3  4
7  8  9 10 11 12 13       5  6  7  8  9 10 11
14 15 16 17 18 19 20     12 13 14 15 16 17 18
21 22 23 24 25 26 27     19 20 21 22 23 24 25
28 29 30                 26 27 28 29 30 31
```

1965 Vehicle Identification Number (VIN) 194x75S100001 to 194x75S123564, total vehicles 23,564, 15,377 convertibles, 8,187 coupes. Plates are located under the glove compartment, riveted to the horizontal brace.
1=Chevrolet
9=Corvette
4=V8 engine
Fourth digit=6 for convertible, 3 for coupe
Sixth digit=model year
S=St. Louis
Six digit sequential serial number starting at 100001

AIR CLEANERS

The 250 and 300HP engines used a dual snorkel style air cleaner. The lid was chrome plated and removable to allow servicing of the original foam element. The base was painted black and had a large, approximately 1", tube extending from the rear to facilitate the hose connection to the breather tube. Cars with air conditioning had pieces attached to the carburetor cutout in order to shift the snorkels away from the radiator hose. The 350 and 365HP air cleaner was basically the same except the base was chrome plated, had a small warm air tube connection, and had cutouts in the rear for air intake replacing the snorkels in the front. The 396 air cleaner was totally different utilizing an open element design that consisted of a top, a base, and a completely exposed paper element. The tops were painted black, although some were reportedly chrome plated. The base contains a large pipe elbow to allow a breather connection to the passenger side valve cover.

The fuel injection air cleaner, and S tube connection to the core support were originally metal. The main canister was mounted to the driver's side inner fender. An elbow was attached to the main canister with three latch type clamps allowing access to the filter element. The uppermost of these three clamps was mounted on top of, not in front of, the fender mounting flange.

AIR CONDITIONING

Extensive changes were required, both small and large, to accommodate the additional equipment necessary for this option. Major additional items included the condenser, mounted in front of the radiator; the evaporator and case, mounted on the passenger side firewall; the dehydrator (drier), mounted on the core support; the compressor, mounted on the engine, passenger side; two additional dash controls; interior ducting consisting of a small outlet above the clock, and two large outlets mounted under the dash at both ends. Adding these parts required different passenger side horn positioning, a new windshield washer reservoir (a bag instead of a plastic jar), battery relocation to the driver's side, moving the headlight motor switch, a longer hood release bracket. and a different trip odometer cable bracket, among other obvious things such as wiring, etc.

The parts themselves evolved into different versions depending on design and manufacturing changes. The compressor mounting was accomplished using a different passenger side exhaust manifold from non air conditioned cars, and the brackets were all cast. The compressor itself had a Frigidaire identification tag on the case with model and serial number, and the rear of the case was dated month/day/year. The month was represented by a letter, Jan=A, Feb=B, etc.

The suction throttling valve, or STV was used to control evaporator pressure from 1963 to 1966, and is mounted on the evaporator case. Both large lines should enter at the top, not top and bottom as seen on POA valves used as replacements.

Original water control valves have a hexagonal portion as part of their construction that was not retained by the later replacements.

Dehydrators (driers) were originally manufactured with a single continuous diameter, later replacements look like an old milk bottle with a smaller diameter at the top.

Replacement or reproduction condensers are more of a universal type having mounting flanges that, unlike the originals, accommodate a number of applications.

The center and right side ducts were the same from 1964 thru 1967. The left side attaches to the side of the steering column brace from 1964 thru 1966. The 1967 left side duct attaches to the underside of the dash brace.

ALTERNATORS

1100693/37 Amp, all except cars equipped with air conditioning and/or transistor ignition
1100694/55 Amp, with air conditioning

1100696/42 Amp, with transistor ignition
1100697/60 Amp, with air conditioning and transistor ignition

All units are stamped with the amperage rating, the model number, and the date. The date is represented year/month/day, with a letter code used for the month, ie A=Jan, B=Feb, etc. Stampings were on the front case. Four blades on the fan should be wider than the others, and the fan should be painted black. The pulley should be cad plated. Cars with 375HP and 365HP without air conditioning used a wide, 3 5/8" diameter pulley with a 1/2" belt groove. The 350 and 365HP with air conditioning used a wide pulley, the diameter was 3 3/64" with 1/2" belt groove. The 350HP without air conditioning and the 425HP used a thin pulley, 3 5/8" diameter with a 1/2" belt groove. Cars with hydraulic lifters, except 350HP, used a narrow 2 13/16" diameter pulley with a 3/8" belt groove.

ANTENNAS

Assembly 3864187. Power antenna manually operated by a switch on the driver's side console. The above the fender assembly consists of a three section mast with an acorn at the top, a chrome nut containing flats, a chrome bezel, and a gasket. All mast sections and the acorn contain a groove around them. The entire motor assembly under the fender is covered in a protective plastic film.

BALANCERS

Narrow, 6" diameter balancer with the pulley bolted directly to it was used on all 250 and 300HP engines. The 350, 365 and 375HP motors used a 1 11/16" wide, 8" diameter balancer with a bolt on pulley. This balancer was dated month/year on the rear face and also contained cooling fins projecting from the inner rear flange. The 396 cubic inch engine used a 1 11/16" wide 8" diameter balancer, undated, without cooling fins. The 396 balancer differed from the 427 balancer in the fact that the rear had a shallow groove in it below the bonded rubber.

BATTERIES

1980558 tar top. Year/month and factory of origin is stamped in the top. Cover above the tar on top should be 2 inches wide. Caps are all black with Delco in yellow on the outside, visible casting line/flashing on the bottom, inside. Case side should say "Delco Original Equipment Power Rated" and should be unpainted (warranty replacements were painted). The characters 559 should be stamped in the side and CAT 558 on top of the battery.

BATTERY CABLES

Both cables are spring style and vinyl covered. The part number is stamped near the terminal end. The negative terminal contains an N and the positive a P, but the P will be covered by a short black sleeve. The positive cable is black, the negative brown.

BELLHOUSINGS

3858403 dated month-day in a small circle, usually unreadable.

BRAKE MASTER CYLINDERS

Cars without disc brakes use leftover 1964 equipment. Cars equipped with standard disc brakes use a single line master cylinder about five inches in diameter. The cap is held in place by two spring steel arms and contains USE ONLY SAE 70R3 BRAKE FLUID stamped around the upper rim. Power brake master cylinders were dual line units and required adaptor fittings to attach the two lines. It used two caps that were a plastic screw in style.

CARBURETORS

Except for Holley, model numbers are embossed on an aluminum tag attached under one of the carburetor top screws. The date is also shown on this tag in the form month/year, using A=Jan, B=Feb, etc., an additional digit indicated factory production information. AFB carburetors have the model number and date stamped on the front of the base, passenger side. The date is in the same form as the aluminum tag.

Holley information was stamped into the front of the air horn. At the top was the Chevrolet part number, the second line was the LIST, or Holley model number, the third line was the date. The date is in the form, year/month/week, each represented by one character. The characters 1 thru 0 represent January thru October, with A and B representing November and December respectively. Be aware that four digit date codes were started later (1970's) in production and are not correct.

All 250 HP with manual transmission used WCFB 3697S, with Powerglide WCFB 3696S. They have the choke mounted at the bottom, and the choke line is attached to the center of the choke cover. A secondary counterweight is used. The main body is cast 0-1465, top 6-1672.

The 300 HP with Powerglide used AFB 3720S, SA or SB. The 300 HP manual transmission cars used AFB 3721S, SA, or SB. The main body is cast 0-1552, top 6-1518 for both.

The 350 and 365 HP used a Holley R2818A. All 396's used Holley R3124A with a dual fuel line connection. The date code should be correct, and there should be visible float adjusting screws and a secondary metering block on all Holleys.

CARPET

Loop carpet, press molded to fit the individual sections and contours of the interior.

COILS

1115202, all 327's without transistor ignition. 1115203 early, 1115207, all 327/350, 365, and 375 horsepower with transistor ignition. 1115210, all 396's which must have transistor ignition. The last 3 digits of the part number appear in raised characters on the case.

CONVERTIBLE TOP

All tops had a rear window attached with heat sealing from the inside. The rear window contained the Vinylite name and trademark, the AS 6 identification, cleaning instructions, and a date code hot stamped into the outside surface. The top frame differed in one respect from 1963 to 1967. The first bow, directly above the driver, was round from 1963 to mid 1965. The pads attached to this bow with screws. The bow used from mid 1965 to 1967 was flat, and the pads were attached to a tack strip.

CRANKCASE VENTS

The rear of the 327 engine block contains a large pipe assembly that is connected to the air cleaner base or FI adaptor with a short piece of rubber hose. A flame arrestor was inserted in this hose on fuel injected cars. The support bracket for the pipe is attached to the outside of the intake manifold on 250 and 300HP, and towards the center of the intake manifold on 350 and 365HP. The pipe on fuel injected engines did not have a support bracket. The oil filler tube had a small pipe that was connected via a piece of rubber hose to the carburetor or fuel injection plenum chamber. Metered orifices were used as opposed to PCV valves in 1965.

The 396 vented the crankcase from both valve covers. The driver's side connected the cover to the carburetor with a small metal pipe. The passenger side connected to the air cleaner with a metal elbow and a piece of rubber hose. Again, as with the 327, metered orifices were used at the carburetor, not PCV valves.

DISTRIBUTORS

The models used with the carbureted engines contain a metal tag that wraps around the neck of the distributor below the cap. This contains the date and model number. All 250, and 300HP cars used model number 1111076. The 350HP without transistor ignition used 1111087. Cars with 350HP and transistor ignition used 1111088. All 365HP cars without transistor ignition used 1111069, those with transistor ignition 1111060. The fuel injection cars without transistor ignition used model 1111070, those with transistor ignition used 1111064. Fuel injection models had a cross shaft access cover. The date and model number were stamped in the cover. All 396's were equipped with transistor ignition and used model number 1111093. All models were dated year/month/day. All models also contained a facility for driving the tachometer cable.

EMBLEMS

Rear Deck

Nose

Side

Side

Side

ENGINE BLOCKS

The 327 cubic inch engine and the 396 cubic inch engine were both used in 1965 with the 396 being introduced very late in the year. The 425 horsepower version of the 396 was the only application. Casting # 3782870, in characters 1/2" high, was used for the small block 327. Smaller characters, and number 3858180 were reportedly used on some 327 equipped cars. The 396 cars used casting # 3855962. The casting number is located on the driver's side rear flange where the bellhousing attaches. The date is located directly opposite on the passenger side rear flange on all 327's except the 3858180 casting. The 3858180 has the date on the same side as the casting number and the date has a two digit year representation. All others are dated month/day/year, with a single digit year representation. The 396 date appears on the passenger side of the block just in front of the starter. The words "HI PERF" were also cast on the passenger side bellhousing flange and just above the oil filter on all 396's. The month is indicated by a consecutive alphabetic character, ie January = A, February = B, etc. A code indicating the assembly location, date, and engine characteristics was stamped on a pad located on the passenger side of the motor. It is just below the cylinder head, next to the water pump. The first character should be F for the Flint assembly plant on 327's and T for Tonawanda on the 396. The numeric characters represent the date, and the last two characters indicate the engine type as shown in the chart that follows. The pad also contained a vehicle serial number. The serial number was on the left on 327's, and on the right on 396's.

- HE = STD 250HP, 4 barrel, manual transmission
- HO = STD 250HP, 4 barrel, automatic transmission
- HI = STD 250HP, 4 barrel, manual, air conditioning
- HQ = STD 250HP, 4 barrel, automatic, air conditioning
- HF = L75 300HP, 4 barrel, high performance, manual
- HP = L75 300HP, 4 barrel, high performance, automatic
- HJ = L75 300HP, 4 barrel, high performance, manual transmission, air conditioning
- HR = L75 300HP, 4 barrel, high performance, automatic, transmission, air conditioning
- HT = L79 350HP, 4 barrel, special high performance with hydraulic lifters, 4 speed
- HU = L79 350HP, 4 barrel, special high performance with hydraulic lifters, air conditioning, 4 speed
- HV = L79 350HP, 4 barrel, special high performance with hydraulic lifters, transistor ignition, 4 speed
- HW = L79 350HP, 4 barrel, special high performance with hydraulic lifters, transistor ignition, air conditioning, 4 speed
- HH = L76 365HP, 4 barrel, special high performance, 4 speed
- HK = L76 365HP, 4 barrel, special high performance, air conditioning, 4 speed

HL = L76 365HP, 4 barrel, special high performance, transistor ignition, 4 speed
HM = L76 365HP, 4 barrel, special high performance, transistor ignition, 4 speed, air conditioning
HG = L84 375HP, fuel injection, 4 speed
HN = L84 375HP, fuel injection, transistor ignition, 4 speed
IF = L78 425HP, 4 barrel, transistor ignition, 4 speed

EXHAUST MANIFOLDS

All 250HP, and 300HP with automatic transmission 3846559 left, 3750556 right, or 3747038 right for the first 1/4 of production, 3747042 for the balance with air conditioning, dated month-day only. The 300HP with manual transmission, 350 and 365HP without air used 3846563 left, 3797902 right, or 3797942 right for 300, 350 and 365HP with air, all undated. The 375HP used 3797942 left, 3797902 right, both undated. The 396 used 3856301 left, and 3856302 right, both dated month-day-year.

FUEL FILTERS

Cars with a WCFB four barrel used a filter mounted in the carburetor just behind the inlet fitting. All other 327's used an external GF 90 filter. The 396's used a similar GF 416 filter. The color of the external filters was black.

FUEL INJECTION

Only one unit was used in 1965 and it used a wide, removable, ribbed plenum cover. The plenum contained a black/aluminum tag showing the part number and a serial number on the driver's side. This serial number has no relationship to any other number. The vehicle serial number was stamped on the rear of the plenum. All chokes were conventional mechanical with a choke tube running from the unit to the driver's side exhaust manifold. A solenoid activated by a micro switch replaced the cranking signal valve for starting purposes. Distributor vacuum was taken directly from the rear side of the plenum. The oil filler tube connected to a restrictor fitting installed in the front of the plenum. One line ran from the main diaphragm to the air meter. One line also ran from the enrichment diaphragm to the plenum. The forward balance tube ran from the fuel meter to the air cleaner adaptor. A splash shield adjustment screw was located on the rear of the main diaphragm cover.

7017380   375HP
                Air meter 7029539 unstamped, (cast 7017248)
                Fuel meter 7017377 unstamped, (cast 7017277)

FUEL PUMPS

The AC logo is cast directly in the top of the screw style pump, and is cast into the side of the crimp style. A four digit Delco number is stamped in the edge of the mounting flange and should be 4657 for 327/250 and 300HP cars without air conditioning. A five digit number, 40083, is stamped in the flange of 327/350, 365 and 375HP cars and cars with air conditioning. The number 40248 is stamped in the flange of 396's. Pumps for 327's are screwed together assemblies using short, 3/4" screws. The 396 pump is a crimp style.

GAUGES

All gauges had glass concave lenses with the exception of the clock which was plastic. The faces were flat with a black background and pale green printed characters. Three tachs were available, all 7000 RPM, with redline variations depending on the motor. All solid lifter equipped cars used a 6500 redline, 350 HP cars used a 6000 redline, and all others used a 5300 redline. Factory replacement tachs usually have an orange and a red warning area where the originals were yellow and orange. Original temp gauges were 240 degree without red and yellow warning areas. Some replacement amp gauges came with 1968 style pointers and some replacement fuel gauges have characters that are too large. Original amp gauges had three terminals in back, the replacements have only two. Oil pressure gauges were 80lb with solid lifter engines, 60lb with all others.

GLASS

All laminated glass was LOF Safety Plate and, along with the logo, was designated as such by AS1 for windshields and AS2 for side and rear glass. Side and rear Plexiglas was AS4. A two character code (month-year) also appeared indicating the date of manufacture. January=N, February=X, March=L, April=G, May=J, June=I, July=U, August=T, September=A, October=Y, November=C, December=V. The letter G indicates 1964, the letter J 1965.

GRILLE

HARDTOP

All tops have reinforcement brackets attached with a button style retainer in the center of the rear window. The window is AS4 Plexiglas. Headliners are made of a fibrous composition material, not vinyl. Headliner colors should match the interior.

HEADS

Casting # 3782461 for all 327's, 3856208 for 396's. Dated month/day/year, with a single digit year representation. For 327's, double digits indicate Tonawanda manufacture and are not correct for Corvette. The month is indicated by a consecutive alphabetic character, ie January = A, February = B, etc. Symbol at the end of the head is a partial indication of the application. The 396 heads must have a provision for attaching the spark plug heat shields.

HORNS

The first four digits of the part number are cast into the horn followed by three stamped digits. The high note is 9000488, the low note 9000487. The horns are also production date stamped year/month/week (1-5). Passenger car horns are assembled with the trumpet indexed at a different location, and the mounting bracket is also different. Mounting brackets were welded to the horn, not bolted.

HORN RELAYS

1115824 was used on all cars. The cover is stamped with Delco Remy and the last three digits of the part number '824', along with '12V', are stamped under the mounting flange.

HUBCAPS

IGNITION SHIELDING

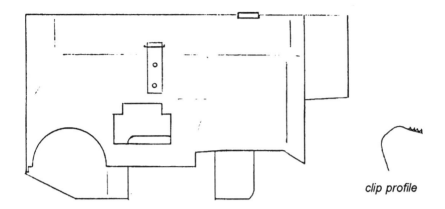

clip profile

396 TOP
The 396 used a top shield with a cover. The 1965 top shield had a front skirt with a semicircular area and a window. No horizontal or vertical shielding was used.

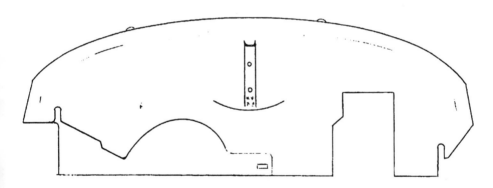

327 TOP
Two plastic rivets hold the insulator. The bulge in the front skirt was higher than the later 1966-1967 top.

VERTICALS
The production right vertical was the same for 1963-1967. The left had a gradual curve to one side, running from top to bottom from 1964-1967.

HORIZONTALS
Two pieces of shielding on each side. All 1964-1965 pieces have the smaller exhaust manifold cut-out.

INTAKE MANIFOLDS

Dated month/day/year. Casting numbers are located on the top surface. Cast iron manifold date codes are also located on the top surface, and are therefore visible with the manifold installed. Correct manifolds should have a single digit year representation. Aluminum manifold date codes are cast into the bottom surface and are not visible unless the manifold is removed from the engine. Cast iron 250HP manifold was 3844457. Cast iron 300HP manifold was 3844459. The 365HP aluminum manifold was 3844461, and the 425HP aluminum manifold was 3866963.

JACK

The original equipment jack is very similar to those supplied for later model Corvettes, and even resembles some from other car models. The correct design has a stamped steel base, arms, and load rest, with a coarse thread jack screw. The load rest is angled to conform to the shape of the Corvette frame, and has the manufacturer's logo, 'A', stamped in it. The size of the base is 4"x6", and overall jack length is about 12 1/2", collapsed from arm to arm. Later jacks are larger. The trunions are the pieces that the jack screw pass through at each end of the jack. These should be U shaped, not solid, and the inside of the U faces the center of the jack until about the end of May. At this time the trunion at the opposite end from the drive nut was turned so that the U faced out. The nut attached to this trunion was always on the outside of the jack, regardless of the U position. At the drive nut end, the load rest support arms are sandwiched between the trunion and the outer arms. These arms contain an oblong stamped reinforcement beginning mid production 1965. The jack shaft is pinned at the drive nut end, and contains a single large thrust washer until the trunion change described above. At this time the large washer was replaced with a combination nylon washer between two steel washers.

KNOBS

LAMPS

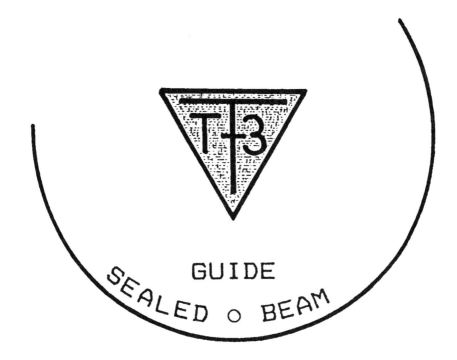

MIRRORS

The outside mirror used a stocky pedestal with the ball stud in the lower 1/3 of the mirror back. A bow tie emblem impression appeared in the back of the mirror. A date code and manufacturers ID appeared on the glass in the form: month/ID/year.

The standard inside mirrors were all basically the same with only minor variations. Some standard mirrors contained rivets on either side of the ball stud, some did not. An optional day- night mirror was also available as part of the comfort and convenience option. A mounting arm attaches to the upper windshield garnish molding and the mirror is attached to the arm with a screw. Standard inside mirrors are dated with the same method as the outside mirrors, day-night mirrors are not dated.

OPTIONS

		Quantity	Price
19437	Base coupe	8,186	$4,321.00
19467	Base convertible	15,376	4,106.00

Code	Option	Qty	Price
Incl.	327, 250HP	1,482	NC
Incl.	3 Speed	404	NC
Incl.	Black folding top	5,565	NC
A01	Tinted glass, all	8,752	16.15
A02	Tinted glass, windshield	7,624	10.80
A31	Power windows	3,809	59.20
C05	White folding top (instead of black)	7,983	NC
C05	Beige folding top (instead of black)	551	NC
C07	Hardtop	6,510	236.75
C07	Hardtop in place of soft top	1,277	NC
C48	Heater delete	39	-100.00
C60	Air conditioning	2,423	421.80
F40	HD Suspension	975	37.70
G81	Positraction rear	19,965	43.50
G91	Highway rear, 3.08 non posi	1,886	2.20
J50	Power brakes	4,044	43.05
J61	Drum brakes (credit)	316	-64.50
K66	Transistor ignition	3,686	75.35
L75	327, 300HP engine	8,358	53.80
L76	327, 365HP engine	5,011	129.15
L78	396, 425HP engine	2,157	292.70
L79	327, 350HP engine	4,716	107.60
L84	327, 375HP engine	771	538.00
M20	Wide ratio 4 speed	8,482	188.30
M20	Close ratio 4 speed	12,625	188.30
M35	Powerglide	2,021	199.10
N03	36 Gallon fuel tank	41	202.30
N11	Off road exhaust	2,468	37.70
N14	Side mounted exhaust	759	134.50
N32	Teakwood steering wheel	2,259	48.45
N36	Telescopic column	3,917	43.05
N40	Power steering	3,236	96.85
P48	Aluminum knock off wheels	1,116	322.80
P91	Nylon blackwalls	168	15.70
P92	Whitewalls	19,300	31.85
T01	Goldline tires	989	50.05
U69	AM/FM radio	22,113	203.40
Z01	Comfort and convenience	15,397	16.15
Z12	Speedometer driven gear	for factory use	
-	Leather seats	2,128	80.70

DEALER AVAILABLE OPTIONS

Compass	7.95
Fire Extinguisher	14.95

Portable spot lamp 8.50
Floor mats 9.25
Locking gas cap 5.75
Luggage carrier 32.50
Tool kit 6.50

PAINT/TRIM

Metal tag with paint and trim codes is located under the glove box.

EXTERIOR	CODE	INTERIOR
Tuxedo Black	AA	Black:Red:Silver:Green:Blue:Saddle:Maroon:White(all combinations)
Nassau Blue	FF	Black:Blue:White/Blue
Glen Green	GG	Black:Saddle:White/Black:Green
Rally Red	UU	Red:Black:White/Red:White/Black
Goldwood Yellow	XX	Black:White/Black
Ermine White	CC	Black:Red:Saddle:Blue:Silver:Green:Maroon:White(all combinations)
Silver Pearl	QQ	Black:Red:Silver
Milano Maroon	MM	Black:Red:Saddle:Wht/Blk:Maroon

These were factory recommended combinations.

A black, white, or beige convertible top was available with any paint/trim combination.

TRIM	VINYL	LEATHER
Saddle	420	421
Blue	414	415
Red	407	408
Black	STD	402
Green	430	431
Maroon	435	436
Sil/Blk	426	427
Wht/Blk	437	438
Wht/Blu	450	451
Wht/Red	443	444

Silver used a black dash and gray carpet.
White and black used a black dash and carpet.
White and blue used a blue dash and carpet.
White and red used a red dash and carpet.

PRODUCTION FIGURES

	MONTHLY	CUMULATIVE
August	227	227
September	1198	1425
October and		

November	1922	3347
December	2407	5754
January	2688	8442
February	2617	11059
March	2877	13936
April	2580	16516
May	2237	18753
June	2463	21216
July	2346	23562
August	2	23564

RADIATORS

All radiators were aluminum with a separately mounted expansion tank. Part number 3155316 was used for all 327's, 3007436 for all 396's. All radiators contained a stamp in the top that showed the part number and date of manufacture represented by a double digit year code and a single alphabetic month code (Jan=A, Feb=B, etc.).

RADIATOR CAPS

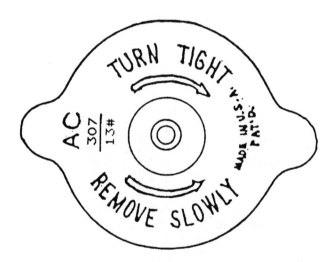

With A/C

RADIOS

Model number 986281, AM/FM. The dial face appears to show AM and FM dark, almost black in color. However, when lit, the face actually shows AM in red and FM in green. A paper tag on the outside of one of the radio covers contains the model number.

REAR AXLE

Code and build date (month/day/year) are located on the rear, lower edge of the carrier.

CODES
AK .... 3.36, Open, all
AL .... 3.08, Positraction, 4 speed
AM .... 3.36, Positraction, manual
AN .... 3.55, Positraction, 4 speed
AO .... 3.70, Positraction, 4 speed
AP .... 4.11, Positraction, 4 speed
AQ .... 4.56, Positraction, 4 speed
AR .... 3.08, Open, 4 speed
AS .... 3.70, Open, 4 speed
AT .... 3.08, Positraction, 396 engine
AU .... 3.36, Positraction, 396 engine
AZ .... 3.55, Positraction, 396 engine
FA .... 3.70, Positraction, 396 engine

FB .... **4.11**, Positraction, 396 engine
FC .... **4.56**, Positraction, 396 engine

SEATS

SHIFTERS

The automatic, three, and four speed shifters used the large 3/4" diameter lever. Early automatic shifters used a somewhat smaller reverse lockout button. The automatic shift pattern remained P R N D L front to back. The four speed lever had a reverse lockout and should not be confused with the later shifter that had a 1/2" longer lever. Lever length should be 6 3/8" from the base to the top without the ball. The later lever also has a greater side bend at the base.

SPARK PLUG WIRES

For 327 carbureted cars the distributor boot is 180 degrees, black. Spark plug boot is 90 degrees, black. For fuel injected cars the distributor and spark plug boot is 90 degrees, black. For 396 cars the distributor boot is 90 degrees, and the spark plug boot is 180 degrees, both boots are black. The 396 spark plug boot is applied over a 135 degree terminal. Wire is 7MM with the imprint PACKARD and an indication that the wire is radio/TV suppression, they are dated quarterly. The date appeared as 3-Q-62, for example, as the third quarter 1962. Radio equipped 396 cars had a woven metal shield over the spark plug wires with an eyelet at one end and a toothed push on clip at the other.

STARTERS

1107320 with 327, 1107352 with 396. The date and model number were stamped directly into the starter housing. The date is shown as year, month, day, and the month appears as an alphabetic character with A=Jan, B=Feb, etc. The solenoid bakelite should contain Delco Remy.

STEERING WHEELS

Standard wheels were plastic simulated wood with metal spokes. The three spokes and center area hub attachment are one piece construction as opposed to the 1967 style which has the center spoke mounted in back of the others. An optional teakwood wheel could be ordered for an additional $48.45. This wheel is the same as the standard wheel except the rim is made entirely of pieces of teakwood.

THERMOSTAT HOUSINGS

All 327 special high performance engines used an angled, cast iron housing, number 3827370. All other engines used a curved, cast iron housing, number 3827369. The 396 required a spacer under the housing for temperature sender clearance.

TIRES

General 'Jet Air', Goodyear 'Power Cushion', Firestone 'Deluxe Champion', Uniroyal 'Laredo', or B.F. Goodrich 'Silvertown' 7.75x15, tubeless, unless equipped with the gold stripe option. The gold stripe tires were offered only as Goodyear 'Power Cushion' or Firestone 'Super Sports', and the stripe width was 3/8" nominal. If whitewall equipped, the width should be 3/4" to 1" nominal. DOT (department of transportation) information did not appear in the sidewall of original tires.

TRANSMISSIONS

The optional four speed was a Muncie for the entire production year. The front bearing retainer was the larger style as opposed to the small 1963 type. The tail housing is a 429, and should have the speedometer gear on the driver's side (late cars may be on the passenger side). The side cover is a 707. Casting dates did not appear on these parts. A build code is stamped on the rear of the case, passenger side, and the vehicle serial number is stamped on the rear of the case, driver's side.

Build information is stamped on the rear of the case, either side, for three speeds. For Powerglide the stamping was on the passenger side of the oil pan. Build information for all transmissions is in the form: location/month/day/shift (D or N).

S=Saginaw (three speed), T=Toledo (Powerglide), P=Muncie (four speed). The three speed must have a shifter mounting position on the tail housing.

VALVE COVERS

All 327 valve cover mounting holes were equally spaced at 8 3/4" top and bottom. The 250 and 300HP engine used a stamped steel cover with a rectangular raised area in the center. The raised area contained a label that read CHEVROLET at the top, 327 in the center, and TURBO-FIRE at the bottom. The sides of the raised area should be triangular.

The remaining 327 engines used a cast aluminum cover with CORVETTE in raised, 1/8" script. A casting flaw should not appear through the 'O' in CORVETTE. All covers had seven fins running the length of the cover. Two notches appeared on the inside lip for intake manifold clearance.

The 396 engines used a painted, stamped steel cover. Except for early production, evidence of spot welds should appear across the top of the cover. This is for the attachment of the oil drippers on the inside. Both covers contain a two wire loom stand in the center. The driver's side should have a flat on the rear corner for power brake booster clearance even if power brakes are not included as an option on the car.

VOLTAGE REGULATORS

1119515 used on all cars. The cover was stamped with Delco Remy on two separate lines and was attached with slotted cap screws. The wire wound resistors underneath should both be the same physical size. The part number and date, along with '12VN' for 12 volt negative ground, are stamped in the mounting flange. The date is shown as year/month, with the month represented by an alphabetic character, ie A=Jan, B=Feb, etc. Beware of the reproductions with unequal size resistors and poor quality on the flange stamping. Some cars with transistor ignition may have used a transistorized regulator, but much of the information to date is inconclusive.

WATER PUMPS

Casting # 3782608, for 250 & 300HP undated. Casting # 3859326, for 350, 365 & 375HP, undated with a small hole to accept the 5/8"NP bypass fitting without using an adaptor. Casting # 3856284 for 396's, dated.

WIPER ARMS & BLADES

All cars used a bright stainless and chrome arm with a bright stainless blade, Trico brand, with rubber insert. The Trico name may not always appear on the parts. Although blueprints show a new blade design for 1965, factory photos clearly show the earlier style still being used. The earlier rubber inserts have a series of raised dots running the length of both sides, the new design has lines. The earlier blade style was a full length center hinge bridge, with two links almost hidden directly below the hinge, and a blade top with a flat cross section. The new style has a peaked cross section for greater strength, and consists of a center bridge with a link at each end.

WIPER MOTORS

5044602 stamped in the armature end housing. The washer pump nozzle should be two pieces, opaque white.

1965

JANUARY
```
S  M  T  W  T  F  S
               1  2
3  4  5  6  7  8  9
10 11 12 13 14 15 16
17 18 19 20 21 22 23
24 25 26 27 28 29 30
31
```

FEBRUARY
```
S  M  T  W  T  F  S
   1  2  3  4  5  6
7  8  9 10 11 12 13
14 15 16 17 18 19 20
21 22 23 24 25 26 27
28
```

MARCH
```
S  M  T  W  T  F  S
   1  2  3  4  5  6
7  8  9 10 11 12 13
14 15 16 17 18 19 20
21 22 23 24 25 26 27
28 29 30 31
```

APRIL
```
S  M  T  W  T  F  S
            1  2  3
4  5  6  7  8  9 10
11 12 13 14 15 16 17
18 19 20 21 22 23 24
25 26 27 28 29 30
```

MAY
```
S  M  T  W  T  F  S
                  1
2  3  4  5  6  7  8
9 10 11 12 13 14 15
16 17 18 19 20 21 22
23 24 25 26 27 28 29
30 31
```

JUNE
```
S  M  T  W  T  F  S
      1  2  3  4  5
6  7  8  9 10 11 12
13 14 15 16 17 18 19
20 21 22 23 24 25 26
27 28 29 30
```

JULY
```
S  M  T  W  T  F  S
            1  2  3
4  5  6  7  8  9 10
11 12 13 14 15 16 17
18 19 20 21 22 23 24
25 26 27 28 29 30 31
```

AUGUST
```
S  M  T  W  T  F  S
1  2  3  4  5  6  7
8  9 10 11 12 13 14
15 16 17 18 19 20 21
22 23 24 25 26 27 28
29 30 31
```

SEPTEMBER
```
S  M  T  W  T  F  S
         1  2  3  4
5  6  7  8  9 10 11
12 13 14 15 16 17 18
19 20 21 22 23 24 25
26 27 28 29 30
```

OCTOBER
```
S  M  T  W  T  F  S
                1  2
3  4  5  6  7  8  9
10 11 12 13 14 15 16
17 18 19 20 21 22 23
24 25 26 27 28 29 30
31
```

NOVEMBER
```
S  M  T  W  T  F  S
   1  2  3  4  5  6
7  8  9 10 11 12 13
14 15 16 17 18 19 20
21 22 23 24 25 26 27
28 29 30
```

DECEMBER
```
S  M  T  W  T  F  S
         1  2  3  4
5  6  7  8  9 10 11
12 13 14 15 16 17 18
19 20 21 22 23 24 25
26 27 28 29 30 31
```

1966

JANUARY
```
S  M  T  W  T  F  S
                  1
2  3  4  5  6  7  8
9 10 11 12 13 14 15
16 17 18 19 20 21 22
23 24 25 26 27 28 29
30 31
```

FEBRUARY
```
S  M  T  W  T  F  S
      1  2  3  4  5
6  7  8  9 10 11 12
13 14 15 16 17 18 19
20 21 22 23 24 25 26
27 28
```

MARCH
```
S  M  T  W  T  F  S
      1  2  3  4  5
6  7  8  9 10 11 12
13 14 15 16 17 18 19
20 21 22 23 24 25 26
27 28 29 30 31
```

APRIL
```
S  M  T  W  T  F  S
                1  2
3  4  5  6  7  8  9
10 11 12 13 14 15 16
17 18 19 20 21 22 23
24 25 26 27 28 29 30
```

MAY
```
S  M  T  W  T  F  S
1  2  3  4  5  6  7
8  9 10 11 12 13 14
15 16 17 18 19 20 21
22 23 24 25 26 27 28
29 30 31
```

JUNE
```
S  M  T  W  T  F  S
         1  2  3  4
5  6  7  8  9 10 11
12 13 14 15 16 17 18
19 20 21 22 23 24 25
26 27 28 29 30
```

JULY
```
S  M  T  W  T  F  S
               1  2
3  4  5  6  7  8  9
10 11 12 13 14 15 16
17 18 19 20 21 22 23
24 25 26 27 28 29 30
31
```

AUGUST
```
S  M  T  W  T  F  S
   1  2  3  4  5  6
7  8  9 10 11 12 13
14 15 16 17 18 19 20
21 22 23 24 25 26 27
28 29 30 31
```

SEPTEMBER
```
S  M  T  W  T  F  S
            1  2  3
4  5  6  7  8  9 10
11 12 13 14 15 16 17
18 19 20 21 22 23 24
25 26 27 28 29 30
```

OCTOBER
```
S  M  T  W  T  F  S
                  1
2  3  4  5  6  7  8
9 10 11 12 13 14 15
16 17 18 19 20 21 22
23 24 25 26 27 28 29
30 31
```

NOVEMBER
```
S  M  T  W  T  F  S
      1  2  3  4  5
6  7  8  9 10 11 12
13 14 15 16 17 18 19
20 21 22 23 24 25 26
27 28 29 30
```

DECEMBER
```
S  M  T  W  T  F  S
            1  2  3
4  5  6  7  8  9 10
11 12 13 14 15 16 17
18 19 20 21 22 23 24
25 26 27 28 29 30 31
```

1966 Vehicle Identification Number (VIN) 194x76S100001 to 194x76S127720, total vehicles 27,720, 17,762 convertibles, 9,958 coupes. Plates are located under the glove compartment, riveted to the horizontal brace.
1=Chevrolet
9=Corvette
4=V8 engine
Fourth digit=6 for convertible, 3 for coupe
Sixth digit=model year
S=St. Louis
Six digit sequential serial number starting at 100001

AIR CLEANERS

All air cleaners were an open element design that consisted of a chrome plated top, a black base and a completely exposed paper element. The base contained a large pipe elbow in order to connect to the engine breather system. This was on the passenger side valve cover for 427's and the rear of the block for 327's. Cars equipped with air injection reactor had an additional base pipe connection for the pump.

AIR CONDITIONING

Extensive changes were required, both small and large, to accommodate the additional equipment necessary for this option. Major additional items included the condenser, mounted in front of the radiator; the evaporator and case, mounted on the passenger side firewall; the dehydrator (drier), mounted on the core support; the compressor, mounted on the engine, passenger side; two additional dash controls; interior ducting consisting of a small outlet above the clock, and two large outlets mounted under the dash at both ends. Adding these parts required different passenger side horn mounting (on the inner fender), a new windshield washer reservoir (a bag instead of a plastic jar), battery relocation to the driver's side, moving the headlight motor switch, a longer hood release bracket. and a different trip odometer cable bracket, among other obvious things such as wiring, etc.

The parts themselves evolved into different versions depending on design and manufacturing changes. Some of the compressor mounting brackets were cast, some stamped. The compressor itself had a Frigidaire identification tag on the case with model and serial number, and the rear of the case was dated month/day/year. The month was represented by a letter, Jan=A, Feb=B, etc.

The suction throttling valve, or STV was used to control evaporator pressure from 1963 to 1966, and is mounted on the evaporator case. Both large lines should enter at the top, not top and bottom as seen on POA valves used as replacements.

Original water control valves have a hexagonal portion as part of their construction that was not retained by the later replacements.

Dehydrators (driers) were originally manufactured with a single continuous diameter, later replacements look like an old milk bottle with a smaller diameter at the top.

Replacement or reproduction condensers are more of a universal type having mounting flanges that, unlike the originals, accommodate a number of applications.

The center and right side ducts were the same from 1964 thru 1967. The left side attaches to the side of the steering column brace from 1964 thru 1966. The 1967 left side duct attaches to the underside of the dash brace.

ALTERNATORS

    1100693/37 Amp, all except cars equipped with air conditioning and/or transistor ignition
    1100750/55 Amp, with air conditioning
    1100696/42 Amp, with transistor ignition
    1100750/60 Amp, with air conditioning and transistor ignition

All units are stamped with the amperage rating, the model number, and the date. The date is represented year/month/day, with a letter code used for the month, ie A=Jan, B=Feb, etc. Stampings were on the front case. Four blades on the fan should be wider than the others, and the fan should be painted black. The pulley should be cad plated on all but the one piece units described below. All 427's with power steering used a wide, 3 5/8" diameter pulley with a 1/2" belt groove. The 350HP 327 and all 427's with air conditioning used a wide pulley, the diameter was 3 3/64" with 1/2" belt groove. The 350HP without air conditioning and the 425HP used a thin pulley, 3 5/8" diameter with a 1/2" belt groove. Cars with 300HP used a narrow one piece fan/pulley assembly with a 3/8" belt groove.

ANTENNAS

Assembly 3864187. Power antenna manually operated by a switch on the driver's side console. The above the fender assembly consists of a three section mast with an acorn at the top, a chrome nut containing flats, a chrome bezel, and a gasket. All mast sections and the acorn contain a groove around them. The entire motor assembly under the fender is covered in a protective plastic film.

BALANCERS

Narrow, 6" diameter balancer with the pulley bolted directly to it was used on all 300HP engines. The 350HP motors used a 1 11/16" wide, 8" diameter balancer with a bolt on pulley. This balancer was dated month/year on the rear face and also contained cooling fins projecting from the inner rear flange. The 427 cubic inch engine with hydraulic lifters used a 1 11/16" wide 7" diameter balancer, undated, without cooling fins. The 427 with solid lifters used an 8" diameter balancer, 1 11/16" wide, undated, without cooling fins.

BATTERIES

1983506 tar top. Year/month and factory of origin is stamped in the top. Cover above the tar on top should be 2 inches wide. Caps are all black with Delco in yellow on the outside, visible casting line/flashing on the bottom, inside. Case side should say "Delco Original Equipment Power Rated" and should be unpainted (warranty replacements were painted). The case should also have hold downs at the bottom. The characters 559 should be stamped in the side and CAT 558 on top of the battery.

BATTERY CABLES

Both cables are spring style and vinyl covered. The part number is stamped near the terminal end. The negative terminal contains an N and the positive a P, but the P will be covered by a short black sleeve. The positive cable is black the negative brown.

BELLHOUSINGS

3858403 aluminum for 327's, 3872444 aluminum for 427's.

BRAKE MASTER CYLINDERS

Cars equipped with standard disc brakes use a single line master cylinder about five inches in diameter. The cap is held in place by two spring steel arms and contains USE ONLY SAE 70R3 BRAKE FLUID stamped around the upper rim. Power brake master cylinders were dual line units and used two caps that were a plastic screw in style.

CARBURETORS

Holley information was stamped into the front of the air horn. At the top was the Chevrolet part number, the second line was the LIST, or Holley model number, the third line was the date. The date is in the form, year/month/week, each represented by one character. The characters 1 thru 0 represent January thru October, with A and B representing November and December respectively. Be aware that four digit date codes were started later (1970's) in production and are not correct.

All 300 and 350 HP without air injection reactor used Holley R3367A, with air injection reactor R3605A. The 390 HP without air injection reactor used Holley R3370A, with air injection reactor R3433A initially then changed to R3606A. The 425 HP used a Holley R3247A (a 3418 may have been used at the end of the year) with a dual fuel line connection. The date code should be correct, and there should be visible float adjusting screws on all Holleys. Only the 425HP carburetor contained a secondary metering block. The choke stove was moved to the intake manifold for all applications.

CARPET

Loop carpet, press molded to fit the individual sections and contours of the interior.

COILS

1115202, all 327's without transistor ignition. 1115232 or 1115262, all 427/390HP without transistor ignition. 1115207, all 327/350HP with transistor ignition. 1115231 or 1115261, all 427's with transistor ignition (some 1115210 early carry over possible). Solid lifter 427's must have transistor ignition. The last 3 digits of the part number appear in raised characters on the case.

CONVERTIBLE TOP

All tops had a rear window attached with heat sealing from the inside. The rear window contained the Vinylite name and trademark, the AS 6 identification, cleaning instructions, and a date code hot stamped into the outside surface. The top frame differed in one respect from 1963 to 1967. The first bow, directly above the driver, was round from 1963 to mid 1965. The pads attached to this bow with screws. The bow used from mid 1965 to 1967 was flat, and the pads were attached to a tack strip.

CRANKCASE VENTS

The rear of the 327 engine block contains a large pipe assembly that is connected to the air cleaner base with a short piece of rubber hose. The support bracket for the pipe is attached to the left side of the intake manifold on 300HP, and towards the center of the intake manifold on 350HP. The oil filler tube had a PCV valve connected to it, and that was connected via a piece of rubber hose to the carburetor.

The 427 vented the crankcase from both valve covers. The driver's side cover contained a PCV valve that was connected to the carburetor with a piece of rubber hose. The passenger side connected to the air cleaner with a metal elbow and a piece of rubber hose.

DISTRIBUTORS

All models used a metal tag that wraps around the neck of the distributor below the cap. This tag contains the date and model number. All 300HP cars used model number 1111153. The 350HP without transistor ignition used 1111156. Cars with 350HP and transistor ignition used 1111157. All 390HP cars without transistor ignition used 1111141, those with transistor ignition 1111142. All cars with 425HP were equipped with transistor ignition and used model number 1111093. All models were dated year/month/day. All models also contained a facility for driving the tachometer cable.

EMBLEMS

Hood & Rear Deck

Side

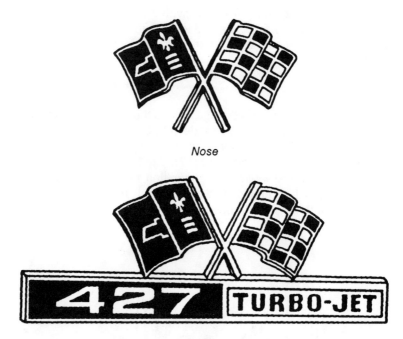

Nose

Side

ENGINE BLOCKS

The 327 cubic inch engine and the 427 cubic inch engine were both used in 1966. The 327 was used for 350 horsepower and below, the 427 for 390 horsepower and above. Casting numbers 3858174 and 3892657 (late) for the 327. 3869942 for the 427. The casting number is located on the driver's side rear flange where the bellhousing attaches. The date is located directly opposite on the passenger side rear flange on all 327's. The 427 date appears on the passenger side of the block just in front of the starter. All are dated month/day/year, with a single digit year representation. The words "HI PERF" were also cast on the passenger side bellhousing flange and just above the oil filter on all 427's. The month is indicated by a consecutive alphabetic character, ie January = A, February = B, etc. A code indicating the assembly location, date, and engine characteristics was stamped on a pad located on the passenger side of the motor. It is just below the cylinder head, next to the water pump. The first character should be F for the Flint assembly plant on 327's and T for Tonawanda on the 427. The numeric characters represent the date, and the last two characters indicate the engine type as shown in the chart that follows. The pad also contained a vehicle serial number. The serial number was on the left on 327's, and on the right on 427's.

HE = STD 300HP, 4 barrel, manual transmission
HO = STD 300HP, 4 barrel, automatic transmission
HH = STD 300HP, 4 barrel, manual transmission, air injection reactor
HR = STD 300HP, 4 barrel, automatic transmission, air injection reactor
HT = L79 350HP, 4 barrel, special high performance, 4 speed
HD = L79 350HP, 4 barrel, special high performance, air injection reactor, 4 speed
HP = L79 350HP, 4 barrel, special high performance with hydraulic lifters, 4 speed, air conditioning, power steering
KH = L79 350HP, 4 barrel, special high performance with hydraulic lifters, air conditioning, 4 speed, air injection reactor, power steering
IL = L36 390HP, 4 barrel, 4 speed
IM = L36 390HP, 4 barrel, air injection reactor, 4 speed
IQ = L36 390HP, 4 barrel, automatic transmission
IR = L36 390HP, 4 barrel, automatic transmission, air injection reactor
IP = L72 425HP, 4 barrel, special high performance, 4 speed
IK = L72 425HP, 4 barrel, special high performance, 4 speed

EXHAUST MANIFOLDS

All 327's without air injection reactor used 3846559 left, 3747042 right. All 327's with A.I.R. used 3872765 left, 3872778 right. Both 327 styles were dated month-day. All 427's used 3880827 left, 3880828 right, dated month- day year.

FUEL FILTERS

All fuel filters were internal, mounted just behind the fuel inlet fitting.

FUEL INJECTION

Not Available

FUEL PUMPS

The AC logo is cast directly in the top of the screw style pump, or in the side of the crimp style. The five digit Delco number is stamped in the edge of the mounting flange and should be 40083 for 350HP or air conditioned 327's, 40366 for 427's. A four digit number, 4657, is stamped in the flange of the 327/300HP without air. Pumps for 327's are screwed together assemblies using short, 3/4" screws. Pumps for 427's are the crimp style.

GAUGES

All gauges had glass concave lenses with the exception of the clock which was plastic. The faces were flat with a black background and pale green printed characters. Three tachs were available, all 7000 RPM, with redline variations depending on the motor. All solid lifter equipped cars used a 6500 redline, 350 HP cars used a 6000 redline, and all others used a 5300 redline. Factory replacement tachs usually have an orange and a red warning area where the originals were yellow and orange. Original temp gauges were 250 degree with red and yellow warning areas. Some replacement amp gauges came with 1968 style pointers and some replacement fuel gauges have characters that are too large. Original amp gauges had three terminals in back, the replacements have only two. Oil pressure gauges were 80lb with solid lifter engines, 60lb with all others. Some original 390HP cars have appeared with 80lb gauges, and some early cars have appeared with 240 degree temp gauges.

GLASS

All laminated glass was LOF Safety Plate and, along with the logo, was designated as such by AS1 for windshields and AS2 for side and rear glass. Side and rear Plexiglas was AS4. A two character code (month-year) also appeared indicating the date of manufacture. January=N, February=X, March=L, April=G, May=J, June=I, July=U, August=T, September=A, October=Y, November=C, December=V. The letter J indicates 1965, the letter A 1966.

GRILLE

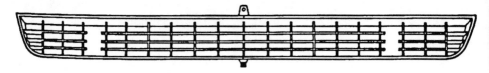

HARDTOP

All tops have reinforcement brackets attached with a button style retainer in the center of the rear window. The window is AS4 Plexiglas. Headliners are made of vinyl not a fibrous composition material.

HEADS

Casting # 3782461 for all 327's, 3872702 all 390HP 427's, and 3873858 all 425HP 427's. Dated month/day/year, with a single digit year representation. For 327's, double digits indicate Tonawanda manufacture and are not correct for Corvette. The month is indicated by a consecutive alphabetic character, ie January = A, February = B, etc. Symbol at the end of the head is a partial indication of the application. The 427 heads must have a provision for attaching the spark plug heat shields.

HORNS

The first four digits of the part number are cast into the horn followed by three stamped digits. The high note is 9000488, the low note 9000487. The horns are also production date stamped year/month/week (1-5). Passenger car horns are assembled with the trumpet indexed at a different location, and the mounting bracket is also different. Mounting brackets were welded to the horn, not bolted.

HORN RELAYS

1115837 was used on all cars. The cover is stamped with Delco Remy and the last three digits of the part number '837', along with '12V', are stamped under the mounting flange.

HUBCAPS

IGNITION SHIELDING

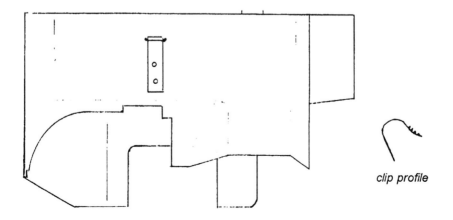

clip profile

427 TOP
The 427 used a top shield with a cover. The 1966 top shield had a front skirt with a curved area and one notch. No horizontal or vertical shielding was used.

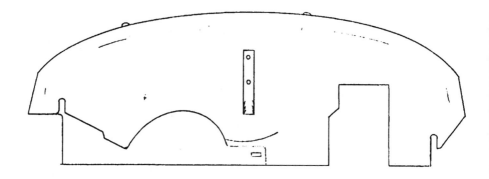

327 TOP
Two plastic rivets hold the insulator. The bulge in the front skirt was lowered for air cleaner clearance.

VERTICALS
The production right vertical was the same for 1963-1967. The left had a gradual curve to one side, running from top to bottom from 1964-1967.

HORIZONTALS
Two pieces of shielding on each side. All 1966-1967 pieces have the larger exhaust manifold cut-out.

INTAKE MANIFOLDS

Dated month/day/year. Casting numbers are located on the top surface. Cast iron manifold date codes are also located on the top surface, and are therefore visible with the manifold installed. Correct manifolds should have a single digit year representation. Aluminum manifold date codes are cast into the bottom surface and are not visible unless the manifold is removed from the engine. Cast iron 300HP manifold was 3872783. Cast iron 390HP manifold was 3866948. The 350HP aluminum manifold was 3890490, and the 425HP aluminum manifold was 3885069.

JACK

The original equipment jack is very similar to those supplied for later model Corvettes, and even resembles some from other car models. The correct design has a stamped steel base, arms, and load rest, with a coarse thread jack screw. The load rest is angled to conform to the shape of the Corvette frame, and has the manufacturer's logo, 'A', stamped in it. The size of the base is 4"x6", and overall jack length is about 12 1/2", collapsed from arm to arm. Later jacks are larger. The trunions are the pieces that the jack screw pass through at each end of the jack. These should be U shaped, not solid, and the inside of the U faces the center of the jack at the drive nut end, and at the opposite end was turned so that the U faced out. The nut was attached to this trunion on the outside of the jack. At the drive nut end, the load rest support arms are sandwiched between the trunion and the outer arms. These arms contain an oblong stamped reinforcement beginning mid production 1965. The jack shaft is pinned at the drive nut end, and contains a thrust washer that is a combination nylon washer between two steel washers.

KNOBS

LAMPS

MIRRORS

The outside mirror used a stocky pedestal with the ball stud in the lower 1/3 of the mirror back. A bow tie emblem impression appeared in the back of the mirror. A date code and manufacturers ID appeared on the glass in the form: month/ID/year.

The standard inside mirrors were all basically the same with only minor variations. Some standard mirrors contained rivets on either side of the ball stud, some did not. A mounting arm attaches to the upper windshield garnish molding and the mirror is attached to the arm with a screw. An optional day- night mirror was supposed to be part of the comfort and convenience group, but this option was canceled and the associated back-up lights were made standard equipment. The day-night mirror was positively not standard equipment in 1966. Standard inside mirrors are dated with the same method as the outside mirrors, day-night mirrors are not dated.

OPTIONS

		Quantity	Price
19437	Base coupe	9,958	$4295.00
19467	Base convertible	17,762	4084.00
Incl.	327, 300HP	9,755	NC
Incl.	3 Speed	564	NC
Incl.	Black folding top	7,259	NC
A01	Tinted glass, all	11,859	15.80
A02	Tinted glass, windshield	9,270	10.55
A31	Power windows	4,562	57.95
A82	Headrests	1,003	42.15
A85	Shoulder harnesses	37	26.35
C05	White folding top (instead of black)	8,789	NC
C05	Beige folding top (instead of black)	411	NC
C07	Hardtop	7,160	231.75
C07	Hardtop in place of soft top	1,303	NC
C48	Heater delete	54	-97.85
C60	Air conditioning	3,520	412.90
F41	HD Suspension	2,705	36.90
G81	Positraction rear	24,056	42.15
J50	Power brakes	5,464	42.15
J56	HD brakes	382	342.30
K19	Air injection reactor	2,380	44.75
K66	Transistor ignition	7,146	73.75
L36	427, 390HP engine	5,116	181.20
L72	427, 425HP engine	5,258	312.85
L79	327, 350HP engine	7,591	105.35

Code	Option	Qty	Price
M20	Wide ratio 4 speed	10,837	184.35
M21	Close ratio 4 speed	13,903	184.35
M22	HD close ratio 4 speed	15	237.00
M35	Powerglide	2,401	194.85
N03	36 Gallon fuel tank	66	198.05
N11	Off road exhaust	2,795	36.90
N14	Side mounted exhaust	3,617	131.65
N32	Teakwood steering wheel	3,941	47.40
N36	Telescopic column	3,670	42.15
N40	Power steering	5,611	94.80
P48	Aluminum knock off wheels	1,194	316.00
P92	Whitewalls	17,969	31.30
T01	Goldline tires	5,557	46.55
U69	AM/FM radio	26,363	199.10
V74	Traffic hazard warning	5,764	11.60
Z01	Comfort and convenience	canceled	
Z12	Speedometer driven gear	for factory use	
-	Leather seats	2,002	79.00

DEALER AVAILABLE OPTIONS

Option	Price
Compass	7.95
Fire Extinguisher	14.95
Portable spot lamp	8.50
Floor mats	9.25
Locking gas cap	5.75
Emergency road kit	10.55
Luggage carrier	32.50
Traffic hazard warning	11.60
Tissue Dispenser	6.95

PAINT/TRIM

Metal tag with paint and trim codes is located under the glove box.

EXTERIOR	CODE	INTERIOR
Tuxedo Black	900	Black:Red:Green:Silver:Bright Blue: Dark Blue:White:Saddle
Nassau Blue	976	Black:Dark Blue:White:Bright Blue
Laguna Blue	978	Black:Bright Blue:Dark Blue
Trophy Blue	980	Black:Bright Blue:Dark Blue
Mossport Green	982	Black:Green
Rally Red	974	Red:Black
Sunfire Yellow	984	Black
Ermine White	972	Black:Red:Saddle:Dark Blue:Silver: Bright Blue:White:Green

Silver Pearl   986   Black:Silver
Milano Maroon   988   Black:Saddle

These were factory recommended combinations.

A black, white, or beige convertible top was available with any paint/trim combination.

TRIM	VINYL	LEATHER
Saddle	420	421
Bright Blue	414	415
Dark Blue	418	419
Red	407	408
Black	STD	402
Green	430	N/A
Silver/Black	426	427
White/Blue	450	N/A

Silver used a black dash and gray carpet.
White used a blue dash and carpet.

PRODUCTION FIGURES

	MONTHLY	CUMULATIVE
September	2031	2031
October	2353	4384
November	2802	7186
December	2706	9892
January	2695	12587
February	2696	15283
March	2808	18091
April	2573	20664
May	2352	23016
June	2453	25469
July	2251	27720

RADIATORS

All 327 radiators were aluminum, part number 3155316, with a separately mounted expansion tank. All contained a stamp in the top that showed the part number and date of manufacture represented by a double digit year code and a single alphabetic month code (Jan=A, Feb=B, etc.). All 427 radiators were copper and did not use an expansion tank.

RADIATOR CAPS

With aluminum radiator

With copper radiator

RADIOS

Model number 986281, AM/FM. The dial face appears to show AM and FM dark, almost black in color. However, when lit, the face actually shows AM and FM in green. A paper tag on the outside of one of the radio covers contains the model number.

REAR AXLE

Code and build date (month/day/year) are located on the rear, lower edge of the carrier.

CODES
AK .... 3.36, Open, all
AL .... 3.08, Positraction, 4 speed
AM .... 3.36, Positraction, manual
AN .... 3.55, Positraction, 4 speed
AO .... 3.70, Positraction, 4 speed
AP .... 4.11, Positraction, 4 speed
AR .... 3.08, Open, 4 speed
AS .... 3.70, Open, 4 speed
AT .... 3.08, Positraction, 427 engine
AU .... 3.36, Positraction, 427 engine
AZ .... 3.55, Positraction, 427 engine
FA .... 3.70, Positraction, 427 engine
FB .... 4.11, Positraction, 427 engine
FC .... 4.56, Positraction, 427 engine

SEATS

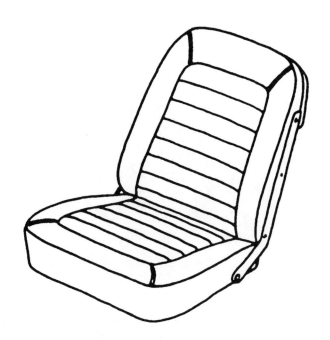

SHIFTERS

The automatic, three, and four speed shifters used the large 3/4" diameter lever. The automatic shift pattern remained P R N D L front to back. The four speed lever had a reverse lockout and should not be confused with the later shifter that had a 1/2" longer lever. Lever length should be 6 3/8" from the base to the top without the ball. The later lever also has a greater side bend at the base.

SPARK PLUG WIRES

For 327 cars the distributor boot is 180 degrees, black. Spark plug boot is 90 degrees, black. For 427 cars the distributor boot is 90 degrees, and the spark plug boot is 180 degrees, both boots are black. The 427 spark plug boot is applied over a 135 degree terminal. Wire is 7MM with the imprint PACKARD and an indication that the wire is radio/TV suppression, they are dated quarterly. The date appeared as 3- Q-62, for example, as the third quarter 1962. Radio equipped 427 cars had a woven metal shield over the spark plug wires with an eyelet at one end and a toothed push on clip at the other.

STARTERS

1107320 all 327's, 1107365 all 427's without M22 transmission, 1107352 all 427's with M22 transmission. The date and model number were stamped directly into the starter housing. The date is shown as year, month, day, and the month appears as an alphabetic character with A=Jan, B=Feb, etc. The solenoid bakelite should contain Delco Remy.

STEERING WHEELS

Standard wheels were plastic simulated wood with metal spokes. The three spokes and center area hub attachment are one piece construction as opposed to the 1967 style which has the center spoke mounted in back of the others. An optional teakwood wheel could be ordered for an additional $47.40. This wheel is the same as the standard wheel except the rim is made entirely of pieces of teakwood.

THERMOSTAT HOUSINGS

All 327/350HP used a curved, cast iron housing, number 3827369. All other engines used an angled aluminum housing, number 3877660.

TIRES

General 'Jet Air', Goodyear 'Power Cushion', Firestone 'Deluxe Champion', Uniroyal 'Laredo', or B.F. Goodrich 'Silvertown' 7.75x15, tubeless, unless equipped with the gold stripe option. The gold stripe tires were offered only as Goodyear 'Power Cushion' or Firestone 'Super Sports', and the stripe width was 3/8" nominal.

If whitewall equipped, the width should be 1/2" to 5/8" nominal. DOT (department of transportation) information did not appear in the sidewall of original tires.

TRANSMISSIONS

The optional four speed was a Muncie for the entire production year. The front bearing retainer was the larger style as opposed to the small 1963 type. The tail housing should have the speedometer gear on the passenger side. The side cover is a 685. Casting dates did not appear on these parts. A build code is stamped on the rear of the case, passenger side, and the vehicle serial number is stamped on the rear of the case, driver's side.

Build information is stamped on the rear of the case, driver's side for three speeds. For Powerglide the stamping was on the passenger side of the oil pan. Build information for all transmissions is in the form: location/month/day/shift (D or N). S=Saginaw (three speed), T=Toledo (Powerglide), P=Muncie (four speed). The three speed must have a shifter mounting position on the tail housing.

VALVE COVERS

All 327 valve cover mounting holes were equally spaced at 8 3/4" top and bottom. The 300HP engine used a stamped steel cover with a rectangular raised area in the center. The raised area contained a label that read 327 at the top, TURBO-FIRE in the center, and 300 HORSEPOWER at the bottom. The sides of the raised area should be triangular.

The remaining 327 engine used a cast aluminum cover with CORVETTE in raised, 1/8" script. A casting flaw should not appear through the 'O' in CORVETTE until late production. All covers had seven fins running the length of the cover. Two notches appeared on the inside lip for intake manifold clearance.

The 427 engines used a painted, stamped steel cover. Evidence of spot welds should appear across the top of the cover. This is for the attachment of the oil drippers on the inside. Both covers contain a two wire loom stand in the center. The driver's side should have a flat on the rear corner for power brake booster clearance even if power brakes are not included as an option on the car. A clearance notch appeared in the front of the passenger side cover during 1966 production, but by the end of the year had been eliminated as unnecessary.

VOLTAGE REGULATORS

1119515 used on all cars. The cover was stamped with Delco Remy on two separate lines and was attached with slotted cap screws. The wire wound resistors underneath should both be the same physical size. The part number and date, along with '12VN' for 12 volt negative ground, are stamped in the mounting flange. The date is shown as year/month, with the month represented by an alphabetic character, ie A=Jan, B=Feb, etc. Beware of the reproductions with unequal size resistors and poor quality on the flange stamping.

WATER PUMPS

Casting # 3782608, for 300HP dated. Casting # 3859326, for 350HP, all except early production were dated, and came with a small hole to accept the 5/8"NP bypass fitting without using an adaptor. Casting # 3856284 for 427's, dated.

WIPER ARMS & BLADES

All cars used a Trico brand, bright stainless and chrome arm with a dull stainless blade and rubber insert until about mid 1966. At this time all parts were given a dull finish. The Trico name may not always appear on the parts. The rubber inserts have a series of lines running the length of both sides. The blade style has a peaked cross section for greater strength, and consists of a center bridge with a link at each end.

WIPER MOTORS

5044602 stamped in the armature end housing. The washer pump nozzle should be two pieces, opaque white.

1966

```
      JANUARY                FEBRUARY
S  M  T  W  T  F  S     S  M  T  W  T  F  S
                  1           1  2  3  4  5
2  3  4  5  6  7  8     6  7  8  9 10 11 12
9 10 11 12 13 14 15    13 14 15 16 17 18 19
16 17 18 19 20 21 22   20 21 22 23 24 25 26
23 24 25 26 27 28 29   27 28
30 31

       MARCH                   APRIL
S  M  T  W  T  F  S     S  M  T  W  T  F  S
         1  2  3  4  5                    1  2
6  7  8  9 10 11 12     3  4  5  6  7  8  9
13 14 15 16 17 18 19   10 11 12 13 14 15 16
20 21 22 23 24 25 26   17 18 19 20 21 22 23
27 28 29 30 31         24 25 26 27 28 29 30

        MAY                    JUNE
S  M  T  W  T  F  S     S  M  T  W  T  F  S
1  2  3  4  5  6  7              1  2  3  4
8  9 10 11 12 13 14     5  6  7  8  9 10 11
15 16 17 18 19 20 21   12 13 14 15 16 17 18
22 23 24 25 26 27 28   19 20 21 22 23 24 25
29 30 31               26 27 28 29 30

        JULY                  AUGUST
S  M  T  W  T  F  S     S  M  T  W  T  F  S
               1  2        1  2  3  4  5  6
3  4  5  6  7  8  9     7  8  9 10 11 12 13
10 11 12 13 14 15 16   14 15 16 17 18 19 20
17 18 19 20 21 22 23   21 22 23 24 25 26 27
24 25 26 27 28 29 30   28 29 30 31
31

     SEPTEMBER                OCTOBER
S  M  T  W  T  F  S     S  M  T  W  T  F  S
            1  2  3                         1
4  5  6  7  8  9 10     2  3  4  5  6  7  8
11 12 13 14 15 16 17    9 10 11 12 13 14 15
18 19 20 21 22 23 24   16 17 18 19 20 21 22
25 26 27 28 29 30      23 24 25 26 27 28 29
                       30 31

      NOVEMBER                DECEMBER
S  M  T  W  T  F  S     S  M  T  W  T  F  S
      1  2  3  4  5              1  2  3
6  7  8  9 10 11 12     4  5  6  7  8  9 10
13 14 15 16 17 18 19   11 12 13 14 15 16 17
20 21 22 23 24 25 26   18 19 20 21 22 23 24
27 28 29 30            25 26 27 28 29 30 31
```

1967

```
      JANUARY                FEBRUARY
S  M  T  W  T  F  S     S  M  T  W  T  F  S
1  2  3  4  5  6  7              1  2  3  4
8  9 10 11 12 13 14     5  6  7  8  9 10 11
15 16 17 18 19 20 21   12 13 14 15 16 17 18
22 23 24 25 26 27 28   19 20 21 22 23 24 25
29 30 31               26 27 28

       MARCH                   APRIL
S  M  T  W  T  F  S     S  M  T  W  T  F  S
         1  2  3  4                       1
5  6  7  8  9 10 11     2  3  4  5  6  7  8
12 13 14 15 16 17 18    9 10 11 12 13 14 15
19 20 21 22 23 24 25   16 17 18 19 20 21 22
26 27 28 29 30 31      23 24 25 26 27 28 29
                       30

        MAY                    JUNE
S  M  T  W  T  F  S     S  M  T  W  T  F  S
   1  2  3  4  5  6                 1  2  3
7  8  9 10 11 12 13     4  5  6  7  8  9 10
14 15 16 17 18 19 20   11 12 13 14 15 16 17
21 22 23 24 25 26 27   18 19 20 21 22 23 24
28 29 30 31            25 26 27 28 29 30

        JULY                  AUGUST
S  M  T  W  T  F  S     S  M  T  W  T  F  S
                  1        1  2  3  4  5
2  3  4  5  6  7  8     6  7  8  9 10 11 12
9 10 11 12 13 14 15    13 14 15 16 17 18 19
16 17 18 19 20 21 22   20 21 22 23 24 25 26
23 24 25 26 27 28 29   27 28 29 30 31
30 31

     SEPTEMBER                OCTOBER
S  M  T  W  T  F  S     S  M  T  W  T  F  S
                  1  2  1  2  3  4  5  6  7
3  4  5  6  7  8  9     8  9 10 11 12 13 14
10 11 12 13 14 15 16   15 16 17 18 19 20 21
17 18 19 20 21 22 23   22 23 24 25 26 27 28
24 25 26 27 28 29 30   29 30 31

      NOVEMBER                DECEMBER
S  M  T  W  T  F  S     S  M  T  W  T  F  S
         1  2  3  4                       1  2
5  6  7  8  9 10 11     3  4  5  6  7  8  9
12 13 14 15 16 17 18   10 11 12 13 14 15 16
19 20 21 22 23 24 25   17 18 19 20 21 22 23
26 27 28 29 30         24 25 26 27 28 29 30
                       31
```

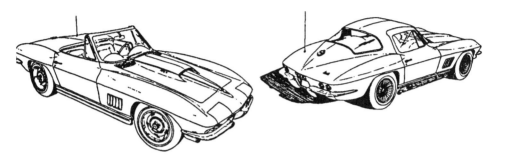

1967 Vehicle Identification Number (VIN) 194x77S100001 to 194x77S122940, total vehicles 22,940, 14,436 convertibles, 8,504 coupes. Plates are located under the glove compartment, riveted to the horizontal brace.
1=Chevrolet
9=Corvette
4=V8 engine
Fourth digit=6 for convertible, 3 for coupe
Sixth digit=model year
S=St. Louis
Six digit sequential serial number starting at 100001

AIR CLEANERS

All air cleaners, except L88 and three two barrel equipped 427's, were a round open element design that consisted of a chrome plated top, a black base and a completely exposed paper element. The base contained a large pipe elbow in order to connect to the engine breather system. This was on the passenger side valve cover for 427's and the rear of the block for 327's. Cars equipped with air injection reactor had an additional base pipe connection for the pump.

The 427 with three two barrels used a triangular shaped open element design that was made up of a black base, a chrome top, and an exposed foam element. The base contained a pipe elbow on the driver's side in order to connect to the valve cover breather. An additional pipe elbow was used on cars with air injection reactor to connect to the pump. L88's used a round base with no provision for a breather connection. The filter element was actually contained in the cowl induction hood and sealed against the base with a piece of foam.

AIR CONDITIONING

Extensive changes were required, both small and large, to accommodate the additional equipment necessary for this option. Major additional items included the condenser, mounted in front of the radiator; the evaporator and case, mounted on the passenger side firewall; the dehydrator (drier), mounted on the core support; the compressor, mounted on the engine, passenger side; two additional dash controls; interior ducting consisting of a small outlet above the clock, and two large outlets mounted under the dash at both ends. Adding these parts required different passenger side horn mounting (on the inner fender), a new windshield washer reservoir (a bag instead of a plastic jar), battery relocation to the driver's side, moving the headlight motor switch, a longer hood release bracket. and a different trip odometer cable bracket, among other obvious things such as wiring, etc.

The parts themselves evolved into different versions depending on design and manufacturing changes. Some of the compressor mounting brackets were cast, some stamped. The compressor itself had a Frigidaire identification tag on the case with model and serial number, and the rear case was dated month/day/year. The month was represented by a letter, Jan=A, Feb=B, etc.

The POA valve was used to control evaporator pressure in 1967, and is mounted on the evaporator case. The large lines should enter at the top and bottom.

Original water control valves have a hexagonal portion as part of their construction that was not retained by the later replacements.

Dehydrators (driers) were originally manufactured with a single continuous diameter, later replacements look like an old milk bottle with a smaller diameter at the top.

Replacement or reproduction condensers are more of a universal type having mounting flanges that, unlike the originals, accommodate a number of applications.

The center and right side ducts were the same from 1964 thru 1967. The 1967 left side duct attaches to the underside of the dash brace. The left side attaches to the side of the steering column brace from 1964 thru 1966.

ALTERNATORS

1100693/37 Amp, all except cars equipped with air conditioning and/or transistor ignition
1100750/55 Amp, with air conditioning
1100696/42 Amp, with transistor ignition
1100750/60 Amp, with air conditioning and transistor ignition

All units are stamped with the amperage rating, the model number, and the date. The date is represented year/month/day, with a letter code used for the month, ie A=Jan, B=Feb, etc. Stampings were on the front case. Four blades on the fan should be wider than the others, and the fan should be painted black. The pulley should be cad plated on all but the one piece units described below. All 427's with power steering used a wide, 3 5/8" diameter pulley with a 1/2" belt groove. The

350HP 327 and all 427's with air conditioning used a wide pulley, the diameter was 3 3/64" with 1/2" belt groove. The 350HP without air conditioning and the solid lifter 427's used a thin pulley, 3 5/8" diameter with a 1/2" belt groove. Cars with 300HP used a narrow one piece fan/pulley assembly with a 3/8" belt groove.

ANTENNAS

The original antenna is a base assembly and a mast assembly with a cap, bezel, and gasket. The correct single mast is 31 1/2" from the top of the 1/4" ball to the bottom of the threads.

BALANCERS

Narrow, 6" diameter balancer with the pulley bolted directly to it was used on all 300HP engines. The 350HP motors used a 1 11/16" wide, 8" diameter balancer with a bolt on pulley. The 427 cubic inch engine with hydraulic lifters used a 1 11/16" wide 7" diameter balancer, undated, without cooling fins. The 427 with solid lifters used an 8" diameter balancer, 1 11/16" wide, undated, without cooling fins.

BATTERIES

1980030. Five push in caps with two half circles pained in red-orange along with one Delco Eye push in cap. E-5000 and R59 are written on the side and Delco Energizer is written on top. Neither are painted, painted writing was found only on warranty replacements.

BATTERY CABLES

Both cables are spring style and vinyl covered. The part number is stamped near the terminal end. The negative terminal contains an N and the positive a P, but the P will be covered by a short black sleeve. The positive cable is black the negative brown.

BELLHOUSINGS

3858403 aluminum for 327's, 3872444 aluminum for 427's.

BRAKE MASTER CYLINDERS

All cars used a dual line cylinder with bleeder screws in the side. The double dome top was held in place by two metal bails. The words USE DELCO BRAKE FLUID and SERVICE WITH DELCO PARTS were stamped in the top.

CARBURETORS

Holley information was stamped into the front of the air horn. At the top was the Chevrolet part number, the second line was the LIST, or Holley model number, the third line was the date. The date is in the form, year/month/week, each represented by one character. The characters 1 thru 0 represent January thru October, with A and B representing November and December respectively. Be aware that four digit date codes were started later (1970's) in production and are not correct.

All 300 and 350 HP without air injection reactor used Holley R3810A, with air injection reactor R3814A. The 390 HP without air injection reactor used Holley R3811A, with air injection reactor R3815A. All 435 HP and the 400 HP with four speed used Holley R3660A in the center and R3659A front and rear. The 400 HP with Powerglide used Holley R3888A in the center and R3659A front and rear. The L88 used a Holley R3418-1. The date code should be correct, and there should be visible float adjusting screws on all Holleys of 400HP or greater. Only the L88 carburetor contained a secondary metering block, and a dual fuel line connection. The choke stove was in the intake manifold for all applications. The L88 was not delivered with a choke.

CARPET

Loop carpet, press molded to fit the individual sections and contours of the interior.

COILS

1115202, all 327's without transistor ignition. 1115207, all 327/350HP with transistor ignition. 1115264, all 427's without transistor ignition. 1115263, all 427's with transistor ignition. Solid lifter 427's must have transistor ignition. The last 3 digits of the part number appear in raised characters on the case.

CONVERTIBLE TOP

All tops had a rear window attached with heat sealing from the inside. The rear window contained the Vinylite name and trademark, the AS 6 identification, cleaning instructions, and a date code hot stamped into the outside surface. The top frame differed in one respect from 1963 to 1967. The first bow, directly above the driver, was round from 1963 to mid 1965. The pads attached to this bow with screws. The bow used from mid 1965 to 1967 was flat, and the pads were attached to a tack strip.

CRANKCASE VENTS

The rear of the 327 engine block contains a large pipe assembly that is connected to the air cleaner base with a short piece of rubber hose. The support bracket for the pipe is attached to the left side of the intake manifold on 300HP, and towards the center of the intake manifold on 350HP. The oil filler tube had a PCV valve connected to it, and that was connected via a piece of rubber hose to the carburetor.

The 427, except for the L88, vented the crankcase from both valve covers. The 390HP driver's side cover contained a PCV valve that was connected to the carburetor with a piece of rubber hose. The passenger side connected to the air cleaner with a metal elbow and a piece of rubber hose. The 400 and 435HP had a flame arrestor inserted in the driver's side valve cover that connected to the air cleaner with a piece of rubber hose. The passenger side cover contained a PCV valve that connected to the intake manifold with a piece of rubber hose. The L88 driver's side valve cover contained a ventilator similar to a road draft tube and the passenger side cover was plugged.

DISTRIBUTORS

All models used a metal tag that wraps around the neck of the distributor below the cap. This tag contains the date and model number. All 300HP cars with manual transmission used model number 1111194, with automatic 1111117. The 350HP without transistor ignition used 1111196. Cars with 350HP and transistor ignition used 1111157. All 390 and 400HP cars without transistor ignition used 1111247, those with transistor ignition 1111294. All cars with 435HP were equipped with transistor ignition and used model number 1111258. Cars with the L88 option used transistor ignition distributor 1111240. All models were dated year/month/day. All models also contained a facility for driving the tachometer cable.

EMBLEMS

Nose

Rear Deck

ENGINE BLOCKS

The 327 cubic inch engine and the 427 cubic inch engine were both used in 1967. The 327 was used for 350 horsepower and below, the 427 for 390 horsepower and above. Casting number 3892657 for the 327. Casting numbers 3869942 (early), 3904351 or 3916321 (very late) for the 427. The casting number is located on the driver's side rear flange where the bellhousing attaches. The date is located directly opposite on the passenger side rear flange on all 327's. The 427 date appears on the passenger side of the block just in front of the starter. All are dated month/day/year, with a single digit year representation. The words "HI PERF" were also cast on the passenger side bellhousing flange and just above the oil filter on all 427's. The month is indicated by a consecutive alphabetic character, ie January = A, February = B, etc. A code indicating the assembly location, date, and engine characteristics was stamped on a pad located on the passenger side of the motor. It is just below the cylinder head, next to the water pump. The first character should be V for the Flint assembly plant on 327's and T for Tonawanda on the 427. The numeric characters represent the date (an I was used as a 1), and the last two characters indicate the engine type as shown in the chart that follows. The pad also contained a vehicle serial number. The serial number was on the left on 327's, and on the right on 427's.

- HE = STD 300HP, 4 barrel, manual transmission
- HO = STD 300HP, 4 barrel, automatic transmission
- HH = STD 300HP, 4 barrel, manual transmission, air injection reactor
- HR = STD 300HP, 4 barrel, automatic transmission, air injection reactor
- HT = L79 350HP, 4 barrel, special high performance, 4 speed
- HD = L79 350HP, 4 barrel, special high performance, air injection reactor, 4 speed
- HP = L79 350HP, 4 barrel, special high performance, power steering, air conditioning, 4 speed
- KH = L79 350HP, 4 barrel, special high performance, air injection reactor, 4 speed, air conditioning
- IL = L36 390HP, 4 barrel, 4 speed
- IM = L36 390HP, 4 barrel, air injection reactor, 4 speed
- IQ = L36 390HP, 4 barrel, automatic transmission
- IR = L36 390HP, 4 barrel, automatic transmission, air injection reactor
- JC = L68 400HP, 3x2 barrels, 4 speed
- JF = L68 400HP, 3x2 barrels, air injection reactor, 4 speed
- JD = L68 400HP, 3x2 barrels, automatic transmission
- JG = L68 400HP, 3x2 barrels, air injection reactor, automatic transmission
- JE = L71 435HP, 3x2 barrels, special high performance, 4 speed
- JA = L71 435HP, 3x2 barrels, special high performance, air injection reactor, 4 speed
- IU = L89 435HP, 3x2 barrels, special high performance, aluminum heads, 4 speed
- JH = L89 435HP, 3x2 barrels, special high performance, aluminum heads, air injection reactor, 4 speed
- IT = L88 Exact horsepower is unavailable, 4 barrel, special engine, heavy duty 4 speed

EXHAUST MANIFOLDS

All 327's without air injection reactor used 3846559 left, 3747042 right. All 327's with A.I.R. used 3872765 left, 3872778 right. Both 327 styles were dated month-day. All 427's used 3880827 left, 3880828 right, dated month-day year.

FUEL FILTERS

All fuel filters were internal, mounted just behind the fuel inlet fitting.

FUEL INJECTION

Not Available

FUEL PUMPS

The AC logo is cast directly in the side of the crimp style pumps or in the top of screwed together pumps. A five digit Delco number is stamped in the edge of the mounting flange and should be 40433 for 327's without air conditioning and 40083 with air conditioning. The 40482 pump was used for all 427's. Pumps are crimped together assemblies except for 40083 which used short 3/4" screws for assembly.

GAUGES

All gauges had concave lenses. The clock, and the optional speed warning tach and speedometer used plastic lenses, all others used glass. The faces were flat with a black background and pale green printed characters. Three tachs were available, all 7000 RPM, with redline variations depending on the motor. All solid lifter equipped cars used a 6500 redline, 350, 390, and 400 HP cars used a 6000 redline, and all others used a 5300 redline. Factory replacement tachs usually have an orange and a red warning area where the originals were yellow and orange. Original temp gauges were 250 degree with red and yellow warning areas. Some replacement amp gauges came with 1968 style pointers and some replacement fuel gauges have characters that are too large. Original amp gauges had three terminals in back, the replacements have only two. Oil pressure gauges were 80lb with solid lifter engines, 60lb with all others.

GLASS

All laminated glass was LOF Safety Plate and, along with the logo, was designated as such by AS1 for windshields and AS2 for side and rear glass. Side and rear Plexiglas was AS4. A two character code (month-year) also appeared indicating the date of manufacture. January=N, February=X, March=L, April=G, May=J, June=I, July=U, August=T, September=A, October=Y, November=C, December=V. The letter A indicates 1966, the letter Z 1967.

GRILLE

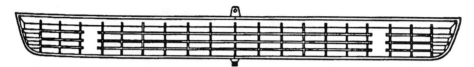

HARDTOP

All tops have reinforcement brackets attached with a button style retainer in the center of the rear window. The window is AS4 Plexiglas. Headliners are made of vinyl not a fibrous composition material.

HEADS

Casting # 3890462 for all 327's. For 390 and 400HP 427's 3904390 early, 3909802 late. For 435HP 427's with cast iron heads 3904391 early, 3919840 late. For 427's with aluminum heads 3904392. Dated month/day/year, with a single digit year representation. For 327's double digits indicate Tonawanda manufacture and are not correct for Corvette. The month is indicated by a consecutive alphabetic character, ie January = A, February = B, etc. Symbol at the end of the head is a partial indication of the application. The 427 heads should not have a provision for attaching spark plug shields.

HORNS

The first four digits of the part number are cast into the horn followed by three stamped digits. The high note is 9000488, the low note 9000487. The horns are also production date stamped year/month/week (1-5). Passenger car horns are assembled with the trumpet indexed at a different location, and the mounting bracket is also different. Mounting brackets were welded to the horn, not bolted.

HORN RELAYS

1115837 was used on all cars. The cover is stamped with Delco Remy and the last three digits of the part number '837', along with '12V', are stamped under the mounting flange.

HUBCAPS

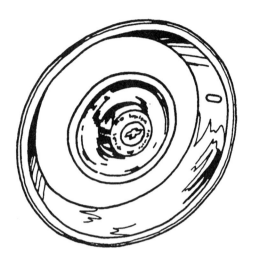

IGNITION SHIELDING

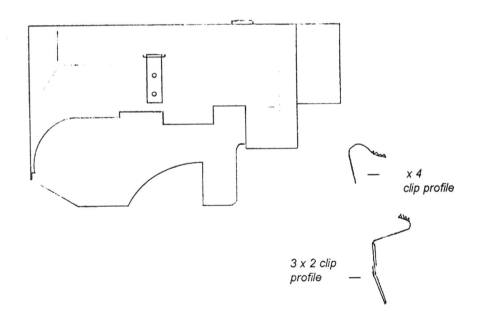

x 4
clip profile

3 x 2 clip
profile

427 TOP

The 427 used a top shield with a cover. The 1967 top shield had a front skirt with a curved area and two notches. The 3x2 cars used a heavier clip that extends to the top of the shield. No horizontal or vertical shielding was used.

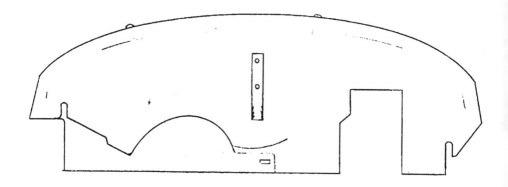

327 TOP

Two plastic rivets hold the insulator. The bulge in the front skirt was lowered for air cleaner clearance.

VERTICALS

The production right vertical was the same for 1963-1967. The left had a gradual curve to one side, running from top to bottom from 1964-1967.

HORIZONTALS

Two pieces of shielding on each side. All 1966-1967 pieces have the larger exhaust manifold cut-out.

INTAKE MANIFOLDS

Dated month/day/year. Casting numbers are located on the top surface. Cast iron manifold date codes are also located on the top surface, and are therefore visible with the manifold installed. Correct manifolds should have a single digit year representation. Aluminum manifold date codes are cast into the bottom surface and are not visible unless the manifold is removed from the engine. Cast iron 300HP manifold was 3872783. Cast iron 390HP manifold was 3866948 early, and 3883948 late. The 350HP aluminum manifold was 3890490. The 400HP aluminum manifold was 3894382, and the 435HP aluminum was 3894374. The L88 used 3885069, also aluminum, sometimes modified to the point where an additional number (3886093) was stamped over, or along with the original number.

JACK

The original equipment jack is very similar to those supplied for later model Corvettes, and even resembles some from other car models. The correct design has a stamped steel base, arms, and load rest, with a coarse thread jack screw. The load rest is angled to conform to the shape of the Corvette frame, and has the manufacturer's logo, 'A', stamped in it. The size of the base is 4"x6", and overall jack length is about 12 1/2", collapsed from arm to arm. Later jacks are larger. The trunions are the pieces that the jack screw pass through at each end of the jack. These should be U shaped, not solid, and the inside of the U faces the center of the jack at the drive nut end, and at the opposite end was turned so that the U faced out. The nut was attached to this trunion on the outside of the jack. At the drive nut end, the load rest support arms are sandwiched between the trunion and the outer arms. These arms contain an oblong stamped reinforcement beginning mid production 1965. The jack shaft is pinned at the drive nut end, and contains a thrust washer that is a combination nylon washer between two steel washers.

KNOBS

LAMPS

MIRRORS

The outside mirror used a stocky pedestal with the ball stud in the lower 1/3 of the mirror back. A bow tie emblem impression appeared in the back of the mirror for about the first 1/3 of production, after that the back was plain. A date code and manufacturers ID appeared on the glass in the form: month/ID/year.

All 1967 inside mirrors are day-night with a gray vinyl covered rim. A mounting arm attaches to the upper windshield garnish molding and the mirror is attached to the arm with a screw. The base of the mounting arm is covered with gray vinyl on all convertibles, and a vinyl color matching the headliner on all coupes. The flipper style changed early in the year from the straight 65-66 design to a longer, curved version.

OPTIONS

		Quantity	Price
19437	Base coupe	8,504	$4388.75
19467	Base convertible	14,436	4240.75
Incl.	327, 300HP	6,858	NC
Incl.	3 Speed	424	NC
Incl.	Black folding top	7,030	NC
A01	Tinted glass, all	11,331	15.80
A02	Tinted glass, windshield	6,558	10.55
A31	Power windows	4,036	57.95
A82	Headrests	1,762	42.15
A85	Shoulder harnesses	1,426	26.35
C05	White folding top (instead of black)	5,900	NC
C05	Blue folding top (instead of black)	611	NC
C07	Hardtop	5,985	231.75
C07	Hardtop in place of soft top	895	NC
C08	Vinyl covering on hardtop	1,966	52.70
C48	Heater delete	35	-97.85
C60	Air conditioning	3,788	412.90
F41	HD Suspension	2,198	36.90
G81	Positraction rear	20,308	42.15
J50	Power brakes	4,760	42.15
J56	HD brakes	267	342.30
K19	Air injection reactor	2,573	44.75
K66	Transistor ignition	5,759	73.75
L36	427, 390HP engine	3,832	200.15
L68	427, 400HP engine	2,101	305.50
L71	427, 435HP engine	3,754	437.10
L79	327, 350HP engine	6,375	105.35
L88	427, HD engine	20	947.90

Code	Description	Qty	Price
L89	427, 435HP with aluminum heads	16	368.65
M20	Wide ratio 4 speed	9,157	184.35
M21	Close ratio 4 speed	11,015	184.35
M22	HD close ratio 4 speed	20	237.00
M35	Powerglide	2,324	194.85
N03	36 Gallon fuel tank	2	198.05
N11	Off road exhaust	2,326	36.90
N14	Side mounted exhaust	4,209	131.65
N36	Telescopic column	2,415	42.15
N40	Power steering	5,747	94.80
N89	Aluminum bolt on wheels	720	263.30
P92	Whitewalls	13,445	31.35
QB1	Redline tires	4,230	46.65
U15	Speed warning speedometer	2,108	10.55
U69	AM/FM radio	22,193	172.75
V48	Maximum coolant protection	336	NC
Z12	Speedometer driven gear	for factory use	
-	Leather seats	1,601	79.00

DEALER AVAILABLE OPTIONS

Item	Price
Compass	7.50
Fire Extinguisher	14.95
Portable spot lamp	8.15
Floor mats	8.90
Locking gas cap	5.52
Emergency road kit	9.95
Luggage carrier	34.50
Ski rack	14.75

PAINT/TRIM

Metal tag with paint and trim codes is located under the glove box.

EXTERIOR	CODE	INTERIOR
Tuxedo Black	900	Black:Red:Green:Bright Blue:Saddle:Teal Blue:White/Blue:White/Black
Marina Blue	976	Black:Bright Blue:White/Blue
Lynndale Blue	977	Black:Teal Blue:White/Black
Elkhart Blue	980	Black:Teal Blue
Goodwood Green	983	Black:Green:Saddle:White/Black
Rally Red	974	Red:Black:White/Black
Sunfire Yellow	984	Black:White/Black
Ermine White	972	Black:Red:Saddle:Teal Blue:Green:Bright Blue:White/Blue:White/Black
Silver Pearl	986	Black:Teal Blue
Marlboro Maroon	988	Black:Saddle:White/Black

A black, white, or blue convertible top was available with any paint/trim combination.

TRIM	VINYL	LEATHER
Saddle	420	421
Bright Blue	414	415
Teal Blue	418	419
Red	407	408
Black	STD	402
Green	430	N/A
White/Blue	450	N/A
White/Black	455	N/A

White/blue used a blue dash and carpet.
White/black used a black dash and carpet.

PRODUCTION FIGURES

	MONTHLY	CUMULATIVE
September	2110	2110
October	575	2685
November	2296	4981
December	2129	7110
January	2355	9465
February	2799	12264
March	3052	15316
April	2079	17395
May	2352	19747
June	2467	22214
July	726	22940

RADIATORS

All 327 and L88 427 radiators were aluminum with a separately mounted expansion tank. Part number for 327's was 3155316, for L88's 3007436. All contained a stamp in the top that showed the part number and date of manufacture represented by a double digit year code and a single alphabetic month code (Jan=A, Feb=B, etc.). All 427 radiators (except L88) were copper and did not use an expansion tank or date tag. Beware of replacements for the 427 that have transmission cooler fittings in the side tank.

RADIATOR CAPS

With aluminum radiator

With copper radiator

RADIOS

Model number 986281, AM/FM. The dial face appears to show AM and FM dark, almost black in color. However, when lit, the face actually shows AM and FM in green. A paper tag on the outside of one of the radio covers contains the model number.

REAR AXLE

Code and build date (month/day/year) are located on the rear, lower edge of the carrier.

CODES
AK .... 3.36, Open, all
AL .... 3.08, Positraction, 4 speed
AM .... 3.36, Positraction, manual
AN .... 3.55, Positraction, 4 speed
AO .... 3.70, Positraction, 4 speed
AP .... 4.11, Positraction, 4 speed
AS .... 3.70, Open, 4 speed
AT .... 3.08, Positraction, 427 engine
AU .... 3.36, Positraction, 427 engine
AZ .... 3.55, Positraction, 427 engine
FA .... 3.70, Positraction, 427 engine
FB .... 4.11, Positraction, 427 engine
FC .... 4.56, Positraction, 427 engine

SEATS

SHIFTERS

The automatic, three, and four speed shifters used the large 3/4" diameter lever. The automatic shift pattern remained P R N D L front to back. The four speed lever had a reverse lockout and should not be confused with the later shifter that had a 1/2" longer lever. Lever length should be 6 3/8" from the base to the top without the ball. The later lever also has a greater side bend at the base.

SPARK PLUG WIRES

For 327 cars the distributor boot is 180 degrees, black. Spark plug boot is 90 degrees, black. For 427 cars the distributor boot is 90 degrees, and the spark plug boot is 180 degrees, both boots are black, except for L88's which are gray. The 427 spark plug boot is applied over a 135 degree terminal. Except for L88's, the wire is 7MM with the imprint PACKARD and an indication that the wire is radio/TV suppression, they are dated quarterly. The date appeared as 3- Q-62, for example, as the third quarter 1962. Radio equipped 427 cars had a woven metal shield over the spark plug wires with an eyelet at both ends. L88 wires were dated month-year and were a dull maroon in color.

STARTERS

1107320 all 327's, 1107365 all 427's without M22 transmission, 1107352 all 427's with M22 transmission. The date and model number were stamped directly into the starter housing. The date is shown as year, month, day, and the month appears as an alphabetic character with A=Jan, B=Feb, etc. The solenoid bakelite should contain Delco Remy.

STEERING WHEELS

All wheels were plastic simulated wood with metal spokes. The three spokes and center area hub attachment has the center spoke mounted in back of the others. The 1963- 1966 style are one piece construction.

THERMOSTAT HOUSINGS

All 327/350HP used a curved, cast iron housing, number 3827369. All other engines, except L88, used an angled aluminum housing, number 3877660.

TIRES

General 'Jet Air', Goodyear 'Power Cushion', Firestone 'Deluxe Champion', Uniroyal 'Laredo', or B.F. Goodrich 'Silvertown' 7.75x15, tubeless, unless equipped with the red stripe option. The red stripe tires were offered only by Uniroyal or Firestone as 'Super Sports', and the stripe width was 3/8" nominal. If whitewall equipped, the width should be 1/2" to 5/8" nominal. DOT (department of transportation) information did not appear in the sidewall of original tires.

TRANSMISSIONS

The optional four speed was a Muncie for the entire production year. The front bearing retainer was the larger style as opposed to the small 1963 type. The tail housing should have the speedometer gear on the passenger side. The side cover is a 685, and has a straight bottom, unlike the curved Borg Warner. Casting dates did not appear on these parts. A build code is stamped on the rear of the case, passenger side, and the vehicle serial number is stamped on the rear of the case, driver's side.

Build information is stamped on the rear of the case, driver's side for three speeds. For Powerglide the stamping was on the passenger side of the oil pan. Build information for all transmissions is in the form: location/year/month/day. S=Saginaw (three speed), T=Toledo (Powerglide), P=Muncie (four speed), month=A=Jan, B=Feb, etc. The three speed must have a shifter mounting position on the tail housing.

VALVE COVERS

All 327 valve cover mounting holes were equally spaced at 8 3/4" top and bottom. The 300HP engine used a stamped steel cover with CHEVROLET in raised script.

The 350HP engine used a cast aluminum cover with CORVETTE in raised, 1/8" script. A casting flaw should appear through the 'O' in CORVETTE. All covers had seven fins running the length of the cover. Two notches appeared on the inside lip for intake manifold clearance.

The 427 engines used a painted, stamped steel cover. Except for early production, evidence of spot welds should appear across the top of the cover. This is for the attachment of the oil drippers on the inside. Both covers contain a two wire loom stand in the center. The driver's side should have a flat on the rear corner for power brake booster clearance even if power brakes are not included as an option on the car.

VOLTAGE REGULATORS

1119515 used on all cars. The cover was stamped with Delco Remy on two separate lines and was attached with slotted cap screws. The wire wound resistors underneath should both be the same physical size. The part number and date, along with '12VN' for 12 volt negative ground, are stamped in the mounting flange. The date is shown as year/month, with the month represented by an alphabetic character, ie A=Jan, B=Feb, etc. Beware of the reproductions with unequal size resistors and poor quality on the flange stamping.

WATER PUMPS

Casting #3782608, for 300HP dated. Casting #3859326, for 350HP, dated with a small hole to accept the 5/8"NP bypass fitting without using an adaptor. Casting #3856284 for 427's, dated.

WIPER ARMS & BLADES

All cars used a Trico brand, dull stainless arm and blade with rubber insert. The Trico name may not always appear on the parts. The rubber inserts have a series of lines running the length of both sides. The blade style has a peaked cross section for greater strength, and consists of a center bridge with a link at each end.

WIPER MOTORS

5044602 stamped in the armature end housing. The washer pump nozzle should be two pieces, opaque white.